Materials Challenges and Testing for Supply of Energy and Resources

Thomas Böllinghaus • Teruo Kishi • Jürgen Lexow
Masaki Kitagawa
Editors

Materials Challenges and Testing for Supply of Energy and Resources

Editors
Thomas Böllinghaus
BAM Bundesanstalt für
Materialforschung und –prüfung
Berlin
Germany

Teruo Kishi
NIMS
Tsukuba
Japan

Jürgen Lexow
BAM Bundesanstalt für
Materialforschung und -prüfung
Berlin
Germany

Masaki Kitagawa
Isobe Mihamaku
Chiba
Japan

"Cover picture: Photo: Pitting corrosion of an autoclave inner surface, BAM Division 6.1 Corrosion and Corrosion Protection, © BAM"
Additional material to this book can be downloaded from http://extras.springer.com
ISBN 978-3-642-23347-0   ISBN 978-3-642-23348-7 (eBook)
DOI 10.1007/978-3-642-23348-7
Springer Heidelberg Dordrecht London New York

Library of Congress Control Number: 2011945285

© Springer-Verlag Berlin Heidelberg 2012
This work is subject to copyright. All rights are reserved, whether the whole or part of the material is concerned, specifically the rights of translation, reprinting, reuse of illustrations, recitation, broadcasting, reproduction on microfilm or in any other way, and storage in data banks. Duplication of this publication or parts thereof is permitted only under the provisions of the German Copyright Law of September 9, 1965, in its current version, and permission for use must always be obtained from Springer. Violations are liable to prosecution under the German Copyright Law.
The use of general descriptive names, registered names, trademarks, etc. in this publication does not imply, even in the absence of a specific statement, that such names are exempt from the relevant protective laws and regulations and therefore free for general use.

Printed on acid-free paper

Springer is part of Springer Science+Business Media (www.springer.com)

Table of Contents

2nd WMRIF Workshop for Young Materials Scientists

Materials challenges for nuclear fission and fusion   **1**

Examination of Dust Particles from Present-Day Controlled Fusion Devices
Elzbieta **Fortuna-Zalesna**, University of Technology, Warsaw, Poland   3

Quantitative microstructural investigation of neutron-irradiated RAFM steel for nuclear fusion applications
Oliver J. **Weiß**, KIT, Karlsruhe, Germany   13

Controlling Welding Residual Stresses by means of Alloy Design
Arne **Kromm**, BAM, Berlin, Germany   23

Degradation Mechanism of Creep Strength Enhanced Ferritic Steels for Power Plants
Kota **Sawada**, NIMS, Tsukuba, Japan   35

Electrochemical studies on pitting corrosion on Cr13 steel exposed to CO_2 and artificial brine with high chloride concentration
Oleksandra **Yevtushenko**, BAM, Berlin, Germany   45

Development of $^{10}B_2O_3$ processing for use as a neutron conversion materials
Lars F. **Voss**, LLNL, Livermore, U.S.A.   55

Materials challenges for water supply   **65**

Water overlayers on Cu(110) studied by van der Waals density Functional
Sheng **Meng**, IPHY CAS, Beijng, China   67

Challenges in conclusive, realistic and system oriented materials testing — 77

Employment of high Resolution RBS to characterize ultrathin transparent electrode in high efficiency GaN based Light Emitting Diode — 79
Grace Huiqi **Wang**, IMRE, Singapore

A possible route to the quantification of piezoresponse force microscopy through correlation with electron backscatter diffraction — 95
Tim L. **Burnett**, NPL, Teddington, United Kingdom

High Resolution Analysis of Tungsten Doped Amorphous Carbon thin Films — 107
Marcin **Rasinski**, University of Technology, Warsaw, Poland

Electron Microscopy Studies on Oxide Dispersion Strengthened Steels — 117
Arup **Dasgupta**, IGCAR, Kalpakkam, India

Fabrication of Probes for In-situ Mapping of Electrocatalytic Activity at the Nanoscale — 129
Andrew J. **Wain**, NPL, Teddington, United Kingdom

Electrochemical Synthesis of Nanostructured Pd-based Catalyst and Its Application to On-Chip Fuel Cells — 143
Satoshi **Tominaka**, NIMS, Tsukuba, Japan

Characterization and Synthesis of PtRu/C Catalysts for Possible Use in Fuel Cells — 153
Eleanor **Fourie**, MINTEK, Randburg, South Africa

Synthesis and investigation of silver-peptide bioconjugates and investigation in their antimicrobial activity — 163
Olga **Golubeva**, ISC RAS, Saint Petersburg, Russia

Characterization of Stabilized Zero Valent Iron Nanoparticles — 173
Lauren F. **Greenlee**, NIST, Boulder, U.S.A.

Combustion Synthesis of Nanoparticles CeO_2 and $Ce_{0.9}Gd_{0.1}O_{1.95}$ — 189
Sumittra **Charojrochkul**, MTEC, Pathumthani, Thailand

Understandings of Solid Particle Impact and Bonding Behaviors in Warm Spray Deposition — 203
Makoto **Watanabe**, NIMS, Tsukuba, Japan

Mechanical properties of innovative metal/ceramic composites based on freeze-cast ceramic performs Siddhartha **Roy**, KIT, Karlsruhe, Germany	213
Mini-Samples Technique in Tensile and Fracture Toughness Tests of Nano-Structured Materials Tomasz **Brynk**, University of Technology, Warsaw, Poland	221
The use of Focused Ion Beam to Build Nanodevices with Graphitic Structures Braulio **Archanjo**, INMETRO, Rio de Janeiro, Brasil	235
Development of compact continuous-wave terahertz (THz) sources by photoconductive mixing Hendrix **Tanoto**, IMRE, Singapore	245
Electrical Impedance Characterization of Cement-Based Materials Supaporn **Wansom**, MTEC, Pathumthani, Thailand	251
On the Use of Indentation Technique as an Effective Method for Characterising starch-based food gels Chaiwut **Gamonpilas**, MTEC, Pathumthani, Thailand	261
Photothermal Radiometry applied in nanoliter melted tellurium alloys Andrea **Cappella**, LNE, Trappes, France	273
Extraction and recovery of scarce elements and minerals	**285**
Biological Treatment of Solid Waste Materials from Copper and Steel Industry Elina **Merta**, VTT, Espoo, Finland	287

Materials challenges for nuclear fission and fusion

Examination of dust particles from present-day controlled fusion devices

E. Fortuna-Zalesna[a], M. Rubel[b], R. Neu[c], M. Rasinski[a], V. Rohde[c],
W. Zielinski[a], M. Andrzejczuk[a], K. J. Kurzydlowski[a], TEXTOR team,
ASDEX Upgrade team

[a] Warsaw University of Technology, Warsaw, Poland

[b] Alfvén Laboratory, KTH, Stockholm, Sweden

[c] Max-Planck-Institut für Plasmaphysik, Garching, Germany

Abstract

Specimens of dust particles collected in the TEXTOR and ASDEX-Upgrade (AUG) tokamaks were examined in order to determine the composition, size and surface and internal structure. The study was performed by means of SEM, TEM, FIB and EDX. The size of particles from both devices varies from hundreds of nananometers to hundreds of micrometers. Dust from TEXTOR contains mainly carbon, boron, silicon, whereas samples from AUG contain also tungsten eroded by plasma from the machine wall.

1. Introduction

Formation of dust particles in magnetic controlled fusion devices is one of the processes arising of plasma-wall interactions (PWI). It is an unwanted consequence of material erosion and migration under regular plasma operation and during off-normal events. Dust formed in large quantities and then re-deposited on diagnostic components may reduce their performance, e.g. optical transmission of diagnostic windows or reflectivity of mirrors. There might be also serious hazards connected with fuel accumulation (so-called tritium inventory) in erosion products. In the case of large quantity of dust (several kg) there may a risk for steam reactions associated with a massive leak of cooling water into the vessel. In present-day machines the amounts of formed dust are small and do not have serious impact for the reactor operation and its safety. However, in future fusion devices operated in steady-state the issue may become important. The access to the in-vessel components of such devices will be extremely limited. This calls for detailed examinations of dust retrieved from the present-day machines.

2. Experimental

Dust samples were collected in March 2008 from TEXTOR and from ASDEX Upgrade, during two openings, in May 2005 and November 2007. TEXTOR samples were collected directly from the machine floor: Inconel liner and graphite tiles shielding components of the dynamic ergodic divertor. The collection was done using a vacuum pump with a series of filters. In ASEDEX-Upgrade (AUG) samples were collected using the adhesive tape technique from 6 different locations on stainless steel parts, see Tab. 1. In the case of AUG only relatively large dust particles (above 200 μm), surrounded by air blisters were selected, removed and examined, because of the collection technique.

Table 1. Location and date of dust particles collection at AUG

Segment Location	Opening	Probe no.
15, below the inner heat shield	02.05.2005	1
5, passive stabilization loop	02.05.2005	2
15, below the inner heat shield	05.11.2007	3
16, from the passive stabilization loop under	05.11.2007	4
11,below divertor baffle	05.11.2007	5
14, electron cyclotron radiation heating bag	05.11.2007	6

The examinations of the dust particles included observations of their size/morphology and examinations of the elemental composition and structure, using scanning electron microscopy (SEM), high resolution scanning electron microscopy (HRSEM), transmission electron microscopy (TEM) and focused ion beam (FIB) techniques.

3. Results

3.1. Dust from TEXTOR

The SEM studies revealed that dust particles differ in elemental composition, morphology and size. Based on their composition they were classified into five following groups: (1) C-O-B mixed, (2) metal-based, (3) ceramic, (4) fine particles (micron and sub-micron sized) and (5) multi-component ones.

The largest group, containing carbon, boron and oxygen, comprises of relatively large particles, from 300 μm to 4 mm in diameter. EDS measurements revealed also small amounts of: Fe, Ni, Cr, Al, Si and Cl. The particles are stratified and cracks are observed on their surface. The particles and/or their sub-layers differ in morphology and chemical composition (C/B/O proportion). Granular, relatively flat, columnar as well as strongly developed structures were observed, as shown in Fig. 1. In some cases, lamellar particles were observed at their surfaces, as shown. These particles are rich in boron and oxygen.

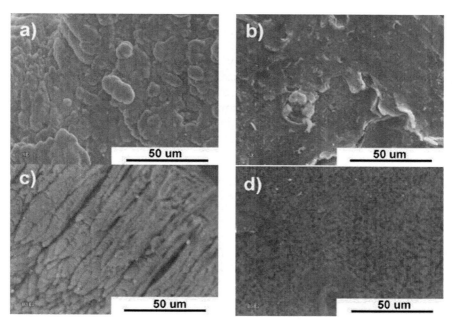

Fig. 1. SEM images illustrating the morphology of dust particles: a) granular, b) relatively flat, c) columnar and d) with high surface roughness

The morphology of metal-based debris (chips with the feed ridge traces) may suggest that they were mechanically removed from the metal parts of the machine (probably during machine opening). The examined particles were either nickel or iron based.

Ceramic debris were also found (100-400 μm). They are very likely small fragments of insulation (alumina, silica, aluminosilicates).

Observations in a BSE mode revealed a large number of fine particles on the surface of larger ones. Their size ranges from several microns to sub-micron. Some of them are fragments of carbon based fakes, a large group, however, is metal based. The examined particles were rich in iron, chromium and nickel. The particles containing ceramic element (Si, Al, O) were also detected.

3.2. Dust from ASDEX-Upgrade

The SEM studies revealed that AUG dust particles differ in their chemical composition, morphology and size. They were classified into four groups: (1) carbon based, (2) C-O-Si particles, (3) metal-based and (4) fine particles (micron and sub-micron sized).

The largest group comprises of relatively large particles, 200 μm to 1.5 mm in diameter, rich in carbon. These particles contain also small amount of oxygen, iron, chromium and silicon. The carbon based particles themselves differ in morphology and can be divided into the following four groups: (a) solid with smooth surface, (b) tissue-like of corrugated surface, (c) porous of anisotropic structures and (d) debris consisting of round, several micron particles, embedded into matrix, as exemplified in Fig. 2.

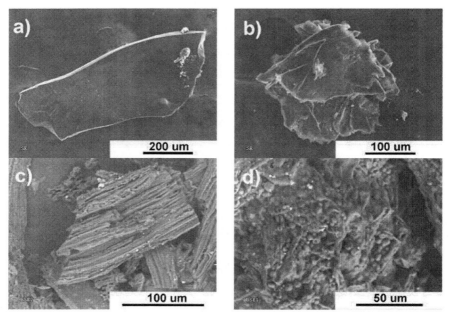

Fig. 2. SEM images of carbon based dust particles: a) flat, b) with corrugated surface, c) anisotropic and d) round particles embedded into matrix

The second largest group comprises relatively large particles, 200 to 500 μm in diameter, containing mainly carbon, oxygen and silicon. The content of silicon and oxygen in these particles amounts to 12-14% and around 30%, respectively. Small amounts of iron and chromium were also detected. Two types of morphologies were observed in this group: uneven surface, grooved particles and smooth ones. The microstructure of grooved particles was examined by TEM. The TEM image shown in Fig. 3 reveals that they contain clusters of nano-particles embedded into a matrix. The EDS analyses reveal the matrix is composed mainly of carbon, whereas the particles seem to be a ternary compounds, containing silicon, carbon and oxygen. Diffraction pattern indicated amorphous character of both matrix and particles.

Several metal-based particles were found, built-up of aluminium, titanium and iron (stainless steel).

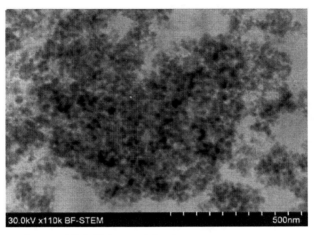

Fig. 3. TEM image of grooved C-O-Si dust particle microstructure

3.3. Fine particles present at "large" ones

Observations in a BSE mode revealed a large number of fine particles on the surface of larger ones, both at AUG and TEXTOR dust. The size of these fine particles varies from well below micron to several microns.

AUG dust fine particles were examined by means of high resolution scanning electron microscopy, Fig. 4. All together 120 particles were examined. It has been found that these particles differ in chemistry. A large group is metal, predominantly tungsten based. Frequently particles rich in iron and chromium, titanium and silver were detected. Fine ceramic particles (Si, Al, O) were also present. Some of the particles are carbon based.

Tungsten fine particles were detected in all locations, both at carbon and metal based large particles. They accounted for 20% of examined particles. Their shape was mainly round, but splashed droplets, flakes and irregular particles were also observed.

Iron based particles were also observed in all locations. Their composition often corresponded to stainless steel (18/8). The observed morphologies of Fe-based particles were: flake/chip, irregular and round.

The carbon based particles were observed mainly at the surface of carbon based large particles. They are small fragments of detached/crushed larger particles.

The size of examined particles was in the range 0.3-5μm.

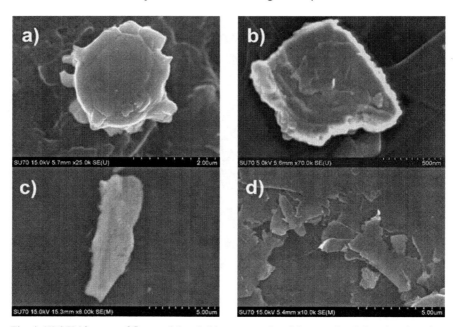

Fig. 4. HRSEM images of fine particles: (a-b) tungsten, c) stainless steel and d) carbon based

4. Discussion

Dust production from eroded plasma-facing components (PFCs) is one of the most critical plasma-wall interaction (PWI) issues. Mobilized dust particles (100nm-100μm) could contain tritium and they also might be a source of other radioactive pollutants if released to the environment. Finally, dust may produce hydrogen after accidental water leak and dust explosions are possible after accidental air ingress.

In principle there are three dust production mechanisms. First is flaking off of deposited layers. Dust particles are released when co-deposited layers tend to disintegrate under the thermal stresses imposed by plasma operation. They can also delaminate by water uptake or air ingress during vents. Second mechanism is a production of droplets during arcing (for example tungsten spheres found in AUG). This happens during normal operation or during ELMs (Edge Localized

Modes) and is important in metal devices. The third term are non normal events like arcs at the structure, work in the vessel or movement of components.

Dust release from tungsten PFCs is mainly due to melt layer movement and droplets ejection. Melting of W samples edges starts above 0.4 MJ/m^2. Even below melting threshold for tungsten cracks developed under repetitive heat loads (100 pulses at 0.8 MJ/m^2) [1], finally also leading to potential dust release.

After long operation periods, the limiters and protected tiles of ICRF antennas of TEXTOR are covered with co-deposited layers containing mainly carbon and hydrogen isotopes (5-16 at.%) but also a significant amount of other species (like B, Si, Ni, Cr, Fe) [2-7]. The presence of other elements is associated with wall conditioning (B from boronization and Si after siliconizaton), erosion of the Inconel liner (Ni, Cr, Fe, Mo as sputtered atoms and metal droplets) and high Z limiters tested as candidate PFCs. Large quantities of oxygen are related to the gettering by Si or B in the machine. However, the majority of oxygen detected in the examined samples is likely to be due to the uptake of oxygen from air during the storage and handling in the laboratory environment.

Because of poor thermal conductivity and adherence, the layers peel off under thermal loads and arcing. In a carbon wall machine the dust particles are predominantly detached flakes of co-deposited layers of the major PFCs.

Different carbon based dust morphologies observed are related to the difference in temperature in various places in TEXTOR. Temperatures in the range of 300-350°C appear in the deposition zone of ALT-II. The temperature 200-250°C is expected at the liner. Temperatures over 2000°C may be reached on the surface of the poloidal limiter. The temperature of PFCs surface influences also fuel content in co-deposits [8].

The area covered by W-PFCs has been increased in AUG steadily since 1999 reaching 85% for the 2005/2006 campaign. In 2007 ASDEX-Upgrade had the first campaign with full tungsten coverage [9,10]. The configuration chosen are tungsten coatings on graphite (SGL carbon R6710 and Schunk F479).

The source of impurities in ASDEX-Upgrade were investigated in the past. Dust collected via filtered vacuum collection technique in July 2000 composed mainly of carbon and constituents of stainless steel - the materials used for most plasma-facing surface and the vacuum vessel [11]. Nearly all particles contained C, Fe, Cr and Mn. The collector probes ex-situ analysis [12] revealed beside deuterium and tungsten, Fe, Ni, Cr and Mn - the components of the stainless steel (SS301) and impurities originating from the covering of the cables containing Ca, K, Si and Ti.

The tungsten erosion is of special interest since tungsten is a candidate material for the ITER divertor. In the inspection performed during the vessel venting of AUG revealed that arcs contributed significantly to the erosion of some W coated PFCs [13]. Especially, arcs were found at the transition and retention modules of the inner divertor. The study revealed a significant fraction of tungsten was not evaporated by the arcs but splashed of as a liquid metal and re-deposited as droplets. Such droplets were detected by SEM observations and profile measurements at the bottom of craters and also at others areas. Typically, the whole thickness of the tungsten coating (3-4 μm) was removed by arcs at these locations as well as small amount of carbon substrate (max 1-2 μm). It was calculated that 2g of tungsten were eroded in the inner divertor. If 10% of the material is eroded as droplets, 10^8 droplets (5μm diameter) are ejected.

The above mentioned observations were a reason for detailed examinations of fine particles. It was confirmed that tungsten particles, submicron and micron size, were present at the surface of the large dust particles, in samples collected from all examined locations.

Part of carbon based-particles are probably small fragments of graphite tiles. The rest originates from disintegrated deposited layers. The structure of mixed C-Si-O particles suggest that a complex and inhomogeneous structures could be expected in co-deposited layers. This should be a motivation for further comprehensive studies of the materials mixing.

The latest results, from 2008 campaign [14], where Si wafer collectors were used to collect dust particle, are in agreement with our results. The submicron particles are predominantly tungsten spheres. The bigger particles often consist of an agglomerate of boron, carbon and tungsten. The amount of carbon in examined dust particles was negligible, which indicates that the formation and subsequent exfoliation of co-deposited a-CH layers is strongly suppressed using W PFCs. However the W droplets may contribute as a source for plasma impurities.

Acknowledgments

This work, supported by European Communities under the contract of Association between EURATOM/IPPLM, EURATOM/IPP, EURATOM/FZJ and EURATOM/VR, was carried out within the EFDA task Force on Plasma Wall Interactions. The view and option expressed herein do not necessarily reflect those of the European Commission.

This work has been a part of the EURATOM-IPPLM Physics Programme founded by European Communities and Polish Ministry of Science and Higher Education under the contracts nos. FU07-CT-2007-00061 and 1170/7PR-EURATOM/2009/7, respectively.

References:
[1] Roth J et al. 2009 J. Nucl. Mater. **390-391** 1
[2] Rubel M et al. 2001 Nucl. Fusin **41** 1087
[3] Rubel M et al. 1991 Mater. Sci. Eng. **A272** 174
[4] Rubel M et al. 1999 J. Nucl. Mater. **266-269** 1185
[5] Rubel M et al.2001 J. Nucl. Mater. **290-293** 473
[6] Rubel M et al 2009, 36th EPS Conference on Plasma Physics, ECA **33**, O4-051
[7] Fortuna E et al. 2007 J. Nucl. Mater. **367-370** 1507
[8] Ivanova D et al. 2009 Phys. Scr. **T138** 014025
[9] Neu R et al. 2009 Phys. Scr. **T138** 014038

[10] Mayer M *et al.* 2009 Phys. Scr. **T138** 014039
[11] Sharpe P *et al.* 2003 *J. Nucl. Mater.* **313-316** 455
[12] Schustereder W *et al.* 2007 *J. Nucl. Mater.* **363-365** 242
[13] Herrmann A *et al.* 2009 *J. Nucl. Mater.* **390-391** 747
[14] Rohde V *et al.* 2009 Phys. Scr. **T138** 014024

Quantitative microstructural investigation of neutron-irradiated RAFM steel for nuclear fusion applications

O. J. Weiß, E. Gaganidze, J. Aktaa

Karlsruhe Institute of Technology, Institute for Materials Research II,

Hermann-von-Helmholtz-Platz 1, 76344 Eggenstein-Leopoldshafen, Germany

Abstract Reduced Activation Ferritic/Martensitic (RAFM) 7-10%Cr-WVTa steels are considered as primary candidate structural materials for in-vessel components of future fusion power plants. These components will be exposed to high neutron and thermo-mechanical loads. Accumulated neutron displacement damage along with transmutation helium generated in the structure materials due to 14.1 MeV fusion neutrons strongly influences the mechanical behavior of the materials. The intention of this work is to evaluate the microstructure of the neutron-irradiated RAFM steel EUROFER97. For this purpose irradiation induced defects like point defect clusters and dislocation loops were identified by transmission electron microscopy (TEM) and quantified in size and volume density. Long term objective is analyzing the influence of the irradiation dose and different neutron fluxes on the evolution of size and density of the defects at irradiation temperature of 300-330 °C.

EUROFER97 samples irradiated to 31.8 dpa were analyzed by TEM. The irradiation was carried out with a flux of 1.8×10^{19} $m^{-2}s^{-1}$ (> 0.1 MeV) in the BOR 60 fast reactor at Joint Stock Company (JSC) "State Scientific Centre Research Institute of Atomic Reactors" (SSC RIAR) in Dimitrovgrad, within the framework of the irradiation program "Associated Reactor Irradiation in BOR 60", which is named ARBOR-1 (Latin for tree).

The quantitative data obtained will be used to correlate the changes in the microstructure to the observed irradiation induced hardening and embrittlement of the material and will serve as an input for models describing this correlation.

1 Introduction

Since fuels are abundant and there is very restricted carbon emission, Nuclear Fusion is a clean way to supply our increasing energy demand. To achieve a fusion reaction the fuel consisting of Deuterium (D) and Tritium (T) – two isotopes of the light element Hydrogen (H) must be heated up to 100 Million degree Celsius. At these extreme temperatures the electrons are separated from the nuclei and the fuel becomes a plasma, which is confined in a magnetic field. Due to thermal energy, the velocity of the ions inside the plasma is high enough to overcome the natural electrostatic repulsion between positively charged atomic nuclei. Thus, when close enough, D-T are fused to heavier Helium (He) atoms, releasing a fast neutron and a total energy of 17.6 MeV per reaction. The uncharged neutrons are not affected by the magnetic field and can leave the plasma to the First Wall/Blanket of the reactor vessel, transferring their energy (14.1 MeV) as heat. In future fusion power plants this heat will be used to produce steam and thereby electricity.

One promising concept of magnetic confinement is the Tokamak with a doughnut-shaped vacuum vessel. This principle is used for the International Thermonuclear Experimental Reactor (ITER) which is currently being built in Cadarache/France [1]. Fig. 1 shows a cut-away view of the ITER Vacuum Vessel showing First Wall/Blanket modules and the Divertor at the bottom. In ITER various types of Test Blanket Modules will be installed. In a later demonstration reactor (DEMO) Breeding Blankets will be used to capture the neutrons, converting their energy to heat. They also allow for neutron multiplication and breeding of Tritium fuel.

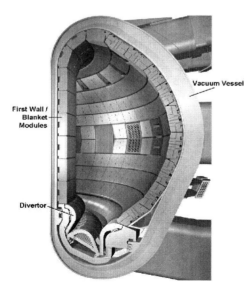

Fig. 1. A cut-away view of the ITER Vacuum Vessel showing the Blanket modules attached to its inner wall and the Divertor at the bottom (taken from www.iter.org).

Function of the Divertor is to pump away He ash and other impurities to maintain a high quality plasma. It consists of cooled structural material (heat sink) and plasma-facing material supported by the former.

Reduced Activation Ferritic/Martensitic (RAMF) steels e.g. EUROFER97 are primary candidate structural materials for the mentioned in-vessel components of future fusion power plants. Their chemical composition is based on the low activation elements Fe, Cr, W, V, Ti, Ta and C. In a fusion reactor the materials will be exposed to high neutron and thermo-mechanical loads. When hitting the material, high-energy neutrons create atomic displacement cascades, and besides a considerable amount of helium is generated by transmutation. The displacement damage accumulates during reactor operation and defects like point-defect clusters and dislocation loops are created. Together with He-bubbles and radiation induced precipitates these defects lead to strong low temperature hardening and embrittlement of the material [2]. Both effects are displayed in Fig. 2 a) and b), respectively.

Aim of this work is to quantify the influence of neutron irradiation on the microstructure of RAFM steels. Since no intensive irradiation facility with a fusion-like neutron spectrum exists, steel samples were irradiated under fission conditions to simulate fusion neutron damage. The samples were then investigated by transmission electron microscopy (TEM). Emphasis is put on measuring the sizes and volume densities of the irradiation-induced defects. These irradiation induced changes in the microstructure should be correlated to changes in mechanical material properties to identify dominant hardening/embrittlement mechanisms in RAFM steels.

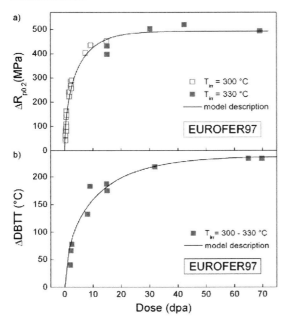

Fig. 2. a) Hardening of EUROFER97: the tensile stress increased with irradiation dose and saturated at about 20 dpa. b) Embrittlement led to an increase of the ductile-to-brittle transition temperature (DBTT) with irradiation dose. The DBTT shift saturated at much higher doses around 60 dpa, indicating the presence of non-hardening embrittlement mechanisms at high doses (from [2]).

2 Experimental

EUROFER97 charpy specimens with the composition 8.91Cr 1.08W 0.48Mn 0.2V 0.14Ta 0.006Ti 0.12C (wt-%, Fe balance) were irradiated to a dose of 31.8 dpa in the BOR 60 fast reactor at Joint Stock Company (JSC) "State Scientific Centre Research Institute of Atomic Reactors" (SSC RIAR) in Dimitrovgrad, within the framework of the irradiation program "Associated Reactor Irradiation in BOR 60", which is named ARBOR-1 (Latin for tree) [3]. The irradiation temperature was 332 °C and the neutron flux 1.8×10^{19} $m^{-2}s^{-1}$ (>0.1 MeV).

The specimens were impact-tested in the material science laboratory of RIAR and thereafter transported to the Fusion Material Laboratory (FML) of Karlsruhe Institute of Technology (KIT). From undeformed ends of the Charpy specimen tested at 138°C slices of 150 μm thickness were cut-off using a cutting wheel. The slices were then thinned by electrolytic polishing in a solution of 20% H_2SO_4 + 80% CH_3OH at room temperature with a Tenupol-5 jet polisher. In order to minimize radioactivity and magnetism, discs of 1 mm diameter including the electron-transparent region were punched out and put in foldable copper nets for examination in the TEM. Modification of the microstructure due to the 1 mm disc punching was not observed.

The TEM investigations were performed at 200 kV using a FEI Tecnai G2 F20 microscope equipped with a post-column GIF Tridiem energy filter (GATAN). The weak-beam dark-field (WBDF) technique [4] was used. To improve the contrast and remove contributions of inelastically scattered electrons in thicker regions of the sample, all images were zero-loss filtered with an energy slit of 15 eV. The diffraction conditions g(7.1g) for g={110} and g(4.1g) for g={200} were chosen, resulting in a diffraction error s_g of approximately 0.2 nm^{-1}.

3 Results

The martensitic microstructure of EUROFER97 irradiated to 32 dpa is shown in Fig. 3 a). Precipitates were located mainly at lath boundaries. Inside the grains the density of defects seems to be quite high. At a higher magnification dislocation loops could be clearly resolved as can be seen in the detailed picture Fig. 3 b). However, due to the bright-field dynamical diffraction condition the image peaks of the defects were very broad giving a complex contrast.

For quantitative analysis weak-beam dark-field imaging was used. The images were thereby confined to the physical sizes of the defects and narrow peaks allowed imaging of small defects down to 1 nm. For the investigation two different grains were chosen. They were tilted to g(7.1g) and g(4.1g) using diffraction vectors of the g={110} and g={200} type, respectively. Two resulting micrographs of

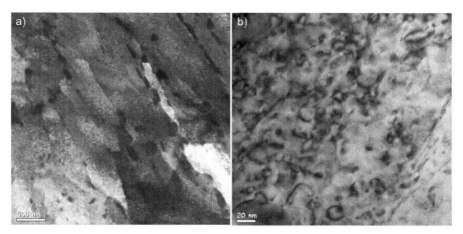

Fig. 3. a) Bright-field image of EUROFER97 irradiated to 32 dpa at 332°C showing the martensitic lath structure. Most of the grain boundaries are decorated with large precipitates. b) Defect microstructure at higher magnification showing dislocation loops and line dislocations.

grain #1 with g={110} are shown in Fig. 4 a) and b). Dislocation loops with diameters of up to 20 nm can be seen, as well as small defect clusters which are visible as white dots and thought to be of interstitial type. The loops generally appear edge on with a strong double arc contrast. To some extend the defects are hard to identify. This is on the one hand because it is not clear whether some white dots are actually small point defect clusters or belong to one of the arcs of larger dislocation loops. On the other hand the identification of the defect is hindered by the detrimental background contrast, which probably stems from surface contamination.

Fig. 4 c) and d) show micrographs of grain #2 with g={200}. Here the background contrast is less pronounced. It appears that most loops are aligned along {211} planes, which is common for bcc systems [6].

After image calibration the defects were marked manually with the line selection tool of the public domain software ImageJ [7]. Counting and sizing were done automatically with the built-in script functions. By this means an area of 1.72 µm² was analyzed. To get the volume density of the defects the foil thickness was measured by convergent beam electron diffraction (CBED) according to Kelly and Allen [8, 9]. For this purpose both investigated grains were tilted to different dynamical two-beam conditions. The recorded CBED patterns were analyzed using a DM-script by V. Hou [10] resulting in average thickness values of 146 nm and 106 nm for grain #1 and #2, respectively. The intensity profile of a {200} diffraction disc is exemplarily shown in Fig. 5.

The size distributions of defect clusters and dislocation loops are given in Fig. 6 a) and b) for grain #1 and #2, respectively. For the shown distributions more than 3600 defects were measured. The maximum loop diameter was 25 nm, while the

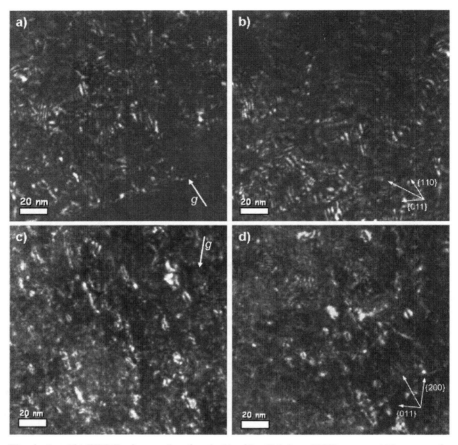

Fig. 4. a) and b) WBDF micrographs of grain #1 with g(7.1g), g={110} near a <111> zone axis; c) and d) WBDF micrographs of grain #2 with g(4.1g), g={200} near a <011> ZA. Most dislocation loops appear edge on with strong double arc contrast and diameters of up to 20 nm (taken from [5]).

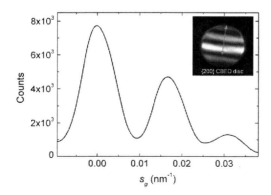

Fig. 5. Image of a {200} CBED disc near a <011> zone axis (inset) and the corresponding intensity profile used for thickness determination (taken from [5]).

Quantitative microstructural investigation of neutron-irradiated RAFM Steel 19

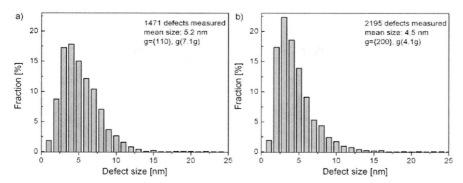

Fig. 6. Size distributions of radiation induced defects in EUROFER97 irradiated to 32 dpa at 332°C. a) Grain #1, the volume density is 1.4×10^{22} m^{-3} b) Grain #2, the volume density is 2.1×10^{22} m^{-3} (taken from [5]).

average value was about 5 nm for both cases. The defect densities were determined to 1.4×10^{22} m^{-3} and 2.1×10^{22} m^{-3}, respectively. To the best of our knowledge quantitative microstructural data of irradiated EUROFER97 was reported only once so far [11]. Similarly to our results, loops with sizes between 5 and 25 nm were detected after irradiation to 16.3 dpa. However, the defect density was about one order of magnitude lower than we measured, mainly because only defects clearly resolvable as dislocation loops were counted [12]. As listed in Table 1, investigations on F82H yielded densities similar to ours already at considerably lower irradiation doses of around 9 dpa. Concerning the dose-dependency it was shown that the defect density first increased with the dose up to a dose between 2.5 and 8.8 dpa and then decreased above [13], which might explain the relatively low defect density in EUROFER97 at 32 dpa. Nevertheless different metallurgical variables like the chemical composition, heat treatments and microstructure of EUROFER97 and F82H might influence defect accumulation during irradiation leading to different defect densities. Furthermore, it is important to notice that due to different orientations with respect to the direction of the electron beam only a fraction of loops is visible in the analyzed micrographs. To get more information about the fraction of invisible loops a section of the sample was imaged using three different diffraction vectors. The resulting micrographs are shown in Fig. 7.

Table 1. Comparison of quantitative microstructural data from different irradiated RAFM steels. Only neutron irradiation with an irradiation temperature T_{irr} in the range of 300-330°C was taken into account (taken from [5]).

Material	Irradiation source	Dose (dpa)	T_{irr} (°C)	Average defect size (nm)	Average defect density (m^{-3})	Reference
EUROFER97	n (BOR60)	31.8	332	4.9	1.75×10^{22}	This work
EUROFER97	n (HFR)	16.3	300	5-25	4.00×10^{21}	[11]
F82H	n (HFR)	8.8	302	5.4	4.50×10^{22}	[13]
F82H	n (HFR)	9.2	310	6.9	2.80×10^{22}	[13]

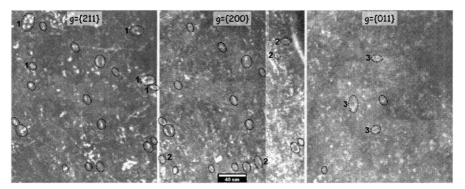

Fig. 7. A section of grain #2 imaged with 3 different diffraction vectors. The defect contrast changes due to different loop inclination after tilting the sample to the next diffraction condition. Loops visible in several images are marked with solid circles. Loops visible in only one of the micrographs are numbered 1-3 and marked with dashed circles. The stripes in the images stem from putting several micrographs together to get exactly the same sample section for each diffraction condition.

Comparing the images it is apparent that the contrast of the defects changed due to different loop inclination after tilting the sample to the next diffraction condition. Most of the loops – marked with closed circles – were visible using the g={211} and g={200} diffraction vectors (partially also with g={011}). Other loops – numbered 1-3 and marked with dashed circles – were only visible in one of the micrographs. To determine the Burgers vectors of the dislocation loops the invisibility criterion g·b=0 was used. Since the diffraction vectors were not exactly known, there is some ambiguity concerning the determination. Tables with all possible g·b combinations were hence used to identify the most probable Burgers vectors: b = <100> in case of the loops marked with '1' and b = ½<111> in all other cases. According to the three micrographs in Fig. 7 roughly 70% of the defects are visible in case of g={211} and g={200} and 19% in case of g={011}. These values are not very reliable, because only a very small area of the sample has been analyzed by this means so far. A systematic comparison of images taken from a much larger area is needed to reliably determine the ratio of loops with <100> and ½<111> type Burgers vectors. However, due to invisibility of defects the overall defect densities could be considerably higher than the reported values. Compared to the irradiation experiments listed in Table 1 the average loop size seems to be rather small for a high-dose irradiation of 32 dpa. Possibly in this case the irradiation temperature was just below the threshold for loop coarsening, which is believed to lie in the range of 330°C [13]. The size distribution of defects could be affected by two points: First, in some cases it was not possible to identify large loops unambiguously because of complicated diffraction and background contrast. Also white dot contrast of the arcs of some larger loops could have been misleadingly counted as small clusters leading to an underestimation of size and density of large loops.

4 Summary and Outlook

The microstructure of EUROFER97 neutron-irradiated to a high-dose of 31.8 dpa was analyzed quantitatively by means of WBDF microscopy. Irradiation induced defects like small point defect clusters and dislocation loops sized between 1 nm and 25 nm have been detected. The average size was 4.9 nm. The average volume density of visible defects was 1.75×10^{22} m^{-3}.

In a subsequent step the results will be compared to quantitative data from samples irradiated to a dose of 15 dpa. The systematic comparison of defects imaged with different diffraction vectors will be extended to larger areas which will allow drawing conclusions about the overall density of defects. The obtained quantitative data will be used to verify the ability of existing physical models for damage cascades to describe the evolution of the density and size of radiation-induced defects with irradiation dose. Our goal is to use the data as an input to model irradiation-induced hardening and embrittlement of EUROFER97.

Acknowledgements This work, supported by the European Communities under the contract of Association between EURATOM and Karlsruhe Institute of Technology (KIT), was carried out within the framework of the European Fusion Development Agreement. The views and opinions expressed herein do not necessarily reflect those of the European Commission. This work was also partly funded by Helmholtz Gemeinschaft e.V. under the grant HRJRG-013.

1 Ikeda, K.: ITER on the road to fusion energy. Nucl. Fusion **50**, 014002 (2010)
2 Gaganidze, E., Schneider, H.-C., Petersen, C., Aktaa, J., Povstyanko, A., Prokhorov, V., Lindau, R., Materna-Morris, E., Möslang, A., Diegele, E., Lässer, R., van der Schaaf, B., Lucon, E.: Mechanical Properties of Reduced Activation Ferritic/Martensitic Steels after High Dose Neutron Irradiation. Proc. of 22nd IAEA Fusion Energy Conference, 13-18 October 2008 Geneva, Switzerland; Paper FT/P2-1
3 Petersen, C., Shamardin, V., Fedoseev, A., Shimansky, G., Efimov, V., Rensman, J.: The ARBOR irradiation project. J. Nucl. Mater. **307-311**, 1655 (2002)
4 Cockayne, D. J. H., Ray, I. L. F., Whelan, M. J.: Investigations of dislocation strain fields using weak beams. Phil. Mag. **20**, 1265 (1969)
5 Weiß, O. J., Gaganidze, E., Aktaa, J.: Quantitative TEM investigations on EUROFER 97 irradiated up to 32 dpa. Submitted to Adv. Sci. Technol., Proceedings of the CIMTEC 2010
6 Byun, T., Hashimoto, N., Farrell, K., Lee, E.: Characteristics of microscopic strain localization in irradiated 316 stainless steels and pure vanadium. J. Nucl. Mater. **349**, 251(2006)
7 ImageJ, Image Processing and Analysis in Java. http://rsbweb.nih.gov/ij/
8 Kelly, P. M., Jostsons, A., Blake, R.. G., Napier, J. G.: The determination of foil thickness by scanning transmission electron microscopy. Phys. Stat. Sol. (a) **31**, 771 (1975)
9 Allen, S. M.: Foil thickness measurements from convergent-beam diffraction patterns. Phil. Mag. A **43**, 325 (1981)
10 Hou, V.: Thickness by CBED. Digital Micrograph(tm) Script Database, http://www.felmi-zfe.tugraz.at/dm_scripts/. Acessed 21 July 2010

11 Klimenkov, M., Materna-Morris, E., Möslang, A.: Characterization of radiation induced defects in EUROFER after neutron irradiation. Submitted to J. Nucl. Mater., Proceedings of the ICFRM-14, Sapporo (2009)
12 Klimenkov, M.: Private communication (2011)
13 Schäublin, R., Victoria, M., Lucas, G., Snead, L., Kirk Jr., M., Elliman, R. (ed.): Identification of Defects In Ferritic/Martensitic Steels Induced by Low Dose Irradiation. MRS Symp. Proc. **650**, R181 (2001)

Controlling Welding Residual Stresses by means of Alloy Design

Arne Kromm

BAM Federal Institute for Materials Research and Testing,

12205 Berlin, Germany

Arne.Kromm@bam.de

Abstract   Residual stresses are a major consequence of many manufacturing processes. In particular, welding residual stresses can affect the processing properties and durability of a component. Especially when welding high strength steels in complex structures, showing high restraint intensities, tensile residual stresses can easily reach the level of the yield strength and cause cracking. Therefore, in this study approaches have been evaluated to control the level and distribution of residual stresses already during the welding process using suitable alloy concepts with reduced phase transformation temperatures. This issue affects primarily the mechanisms between transformation temperature, transformation kinetics and resulting residual stresses. The present work includes discussion and approaches, how these phenomena can be evaluated using an appropriate in-situ monitoring. For that purpose, investigations using high energy synchrotron diffraction have been carried out. Initially, diffraction analyses have been made during simple thermal cycles, focusing on the observation of phase transformation kinetics, i.e. transformation temperatures and temperature dependent phase formation and contents. Furthermore, welding experiments incorporating an external shrinkage restraint show how the stress formation during cooling is influenced by superimposed mechanical and thermal impacts compared to conventional filler material. The results demonstrate that residual stresses can be effectively controlled by means of an adjusted alloy design. Prior to experimental results a review is presented including the mechanisms of stress reduction due to phase transformation during welding. Beside this the progress in designing special alloys for controlling residual stresses is reflected.

Introduction

The residual stress state of welded constructions is essentially influenced by the interaction of mechanical, thermal and material factors. The phase transformation during welding is determined by the chemical composition and by the cooling rate. Depending on these two variables different transformation temperatures result. On the other hand microstructures and therefore mechanical properties are varying. This influences the residual stress formation in relatively complex interaction, as discussed in detail by Wohlfahrt [1]. Regarding the mechanisms of phase transformations on the welding residual stresses, mostly the work of Jones and Alberry [2] is referred. However, first fundamental studies on the effect of the M_s-temperatures on the resulting residual stresses after heat treatments were already carried out by Bühler and Scheil [3] using iron-base materials with varying nickel contents. According to the authors, the residual stresses after cooling are the results of two competing processes: the hindered shrinkage on the one hand and on the other hand, the volume expansion as a result of the phase transformation. Additionally impacts of quenching are significant. The results showed that the level of residual stresses is mainly defined by the M_s-temperature. Another relevant factor is whether the M_f-temperature is located above or below ambient temperature. Jones und Alberry [2] have confirmed these findings in principle for welding. According to Bhadeshia [4] the residual stress reduction due to phase transformation can be attributed to the following reasons: Due to the difference in thermal expansion coefficients of austenite and ferrite, the volume expansion at low temperatures is larger. Furthermore, the fcc lattice and therefore high stress compensation by the lower yield strength of austenite prevents the development of high stresses. In contrast, in the bcc-lattice lower shrinkage compensation is available, leading to comparatively high stress gradients. For these reasons, regarding the reduction of welding residual stress by means of phase transformation it is essential to realize lowest possible transformation temperatures by corresponding alloy concepts. Another factor is that largest possible deformation and thus high transformation plasticity has to be achieved. This means that the formation of martensite is desired.

Based on this knowledge up to now three different approaches of alloys with special adjusted martensite start temperatures appeared in the literature. Based on very first investigations by Murata et al. [5] the research group around Ohta [6-7] seized this approach and has designed an iron-based alloy with the main alloying elements nickel and chromium. This chemical composition leads to a martensite start temperature of $M_s = 180$ °C. Shiga et al. [8] recently proofed the existance of compressive residual stresses when welding this kind of LTT alloy. Martinez-Diez [9] developed an alternative LTT composition substituting nickel by manganese. Another approach using solely nickel as main alloying element was recently proposed in [10]. Up to now different approaches are being pursued to realize LTT alloys. The targeted M_s-temperatures could be successfully adjusted by different chemical compositions. First residual stress measurements have demonstrated the effectiveness of these alloys with regard to the welding residual stresses. However, level and distribution of welding residual stresses in different joint configurations are largely unexplored. The use of neutron radiation, which allows measurements in the weld volume, however, demonstrates the importance attached to the clarification of these questions. Further research is needed to understand the mechanisms of residual stress formation in interaction with phase transformation, shrinkage and quenching effects especially with this new group of materials. Less importance has been attached to the austenitic phase of these materials. It should be noted that due to various factors, such as micro-segregation and incomplete martensite formation, considerably austenite contents can occur, which should influence the macro residual stress state. The following investigations show how especially in-situ methods are appropriate to characterize the transformation behavior of a specific LTT alloy.

Material

For this study a LTT alloy was selected with the chemical composition shown in table 1. The alloy was already proven to be suitable for residual stress control during welding [11]. For reasons of comparison a high strength filler material with similar strength was chosen as reference material. The chemical composition and the mechanical properties are given in table 2 and table 3. Since the intended applications of LTT alloys are high strength joints the high strength structural steel S690Q was used as base material (see table 4).

Table 1. Chemical compositon of the LTT alloy (wt.%)

C	Cr	Ni	Si	Mn	Fe
0.04	10	8	0.4	0.7	balance

Table 2. Mechanical Properties of the high strength reference filler

R_m in MPa	$R_{p0.2}$ in MPa	Elongation A_5	Impact energy
475	596	25	≥ 40 J

Table 3. Chemical composition of the high strength reference filler (wt.%)

C	Ni	Si	Mn	P	S	N	O	Fe
0.05	3.7	0.2	0.7	0.008	0.007	0.01	0.008	balance

Table 4. Chemical composition of the high strength base material (wt.%)

C	Cr	Ni	Si	Mn	P	S	Mo	Nb	V	B	Fe
0.116	0.498	0.481	0.402	1.52	0.017	<0.001	0.111	<0.005	0.054	0.005	balance

Experimental

The experimental investigations of this work are subdivided into in-situ diffraction measurements of the transformation behavior of the undiluted LTT alloy during certain thermal cycles and online stress measurements during a real welding process under defined restraint conditions.

In-situ Diffraction

Diffraction measurements were performed at the Helmholtz Zentrum Berlin using high energy synchrotron radiation provided on the materials science beamline EDDI [12]. To investigate the transformation behavior during cooling a domed heating stage (Type Anton Paar DHS 1100) specially developed for use on diffractometers was applied. Penny shape specimen with a diameter of 10 mm and a thickness of 30 μm were produced from undiluted LTT alloy. The specimens were heated up continuously to 1100°C and immediately cooled down to room temperature by a cooling rate of 250 K/min using nitrogen as shielding gas. This rather slow cooling rate ensured accurate capturing of the M_s-temperature. The LTT alloy shows martensite transformation even when air quenching is applied. Diffraction spectra were recorded every 5 s during the thermal cycle. The use of energy dispersive diffraction methods allowed for measuring large diffraction patterns containing information of all crystalline phases appearing in the material. The identification and quantification of martensite and also austenite at specific temperatures could be reached within one measurement that way. Table 5 shows the measuring and evaluation parameters.

Table 5. Measuring and evaluation parameters applied in in-situ diffraction measurement

Primary beam	1x1 mm²
Secondary side optics	Double slit system (equatorial × axial) 0.03 x 5 mm²
Diffraction angle	$2\theta = 6°$
Measuring mode	Reflection
Acquisition time	5 s / spectrum
Calibration	Tungsten powder

To evaluate the kinetic of the transformation during cooling the content of autenite and martensite has to be known. Following [13] this contents can be determined taking into account the ratios of the integrated intensities of the individual Bragg reflections of both phases weighted by a proportionality factor, which can be calculated from basic principles (eq. 1).

$$V_\alpha = \cfrac{1}{\left[1 + \left(\cfrac{R_{\alpha(hkl)}}{R_{\gamma(hkl)}} \cfrac{I_{\gamma(hkl)}}{I_{\alpha(hkl)}}\right)\right]} \quad (1)$$

V_α – volume fraction ferrite/martensite
$I_\alpha(hkl)$ – integrated intensity of individual ferrite/martensit reflection
$I_\gamma(hkl)$ – integrated intensity of individual austenite reflection
$R_\alpha(hkl)$ – proportionality factor of individual ferrite/martensite reflection
$R_\gamma(hkl)$ – proportionality factor of individual austenite reflection

Online Stress Measurements

The online stress measurements during welding were performed using a special welding test device for medium scale weldability tests shown in Fig.1. The setup allows for applying defined restraint intensities by preventing free shrinkage of the specimens during welding. The specimen is fixed using two hydraulic clamps, one located in the stiff frame and the second one in a movable traverse. The reaction force is determined via strain gauges, applied at the frame. Measuring of the elastic straining of the frame during welding and cooling allows determination of the accumulating forces.

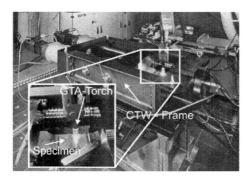

Fig. 1. Controlled Tensile Weldability (CTW) test for online measurement of reaction forces during welding and subsequent cooling

Specimens offering V-notch joints were welded using the LTT alloy in two layers taking the high strength steel S690Q (table 4) as base material. The heat input was determined to 1 kJ/mm. The plate thickness was 6 mm. Manual metal arc welding (MMAW) was conducted on a length of 100 mm. During welding the temperature was captured by immersing a thermocouple into the liquid weld pool. The clamping length was predefined to 170 mm. The restraint intensity could be estimated to 7.4 kN/mm·mm. For reasons of comparison a conventional high strength filler material (table 2 and table 3) was additionally applied under similar conditions. Metallographic examination was conducted after welding.

Results and Discussion

In-situ Diffraction

In order to visualize the austenite to martensite phase transformations during cooling, the measured diffraction spectra are compiled as 2D-density plots in Fig.2. The energy is plotted versus the temperature. High diffraction intensities are depicted by white color. The diffraction lines in the diagram at 57 keV, 58 keV, 66 keV, 83 keV, 93 keV and 101 keV can be assigned to the 111γ, 110α, 200γ, 200α, 220γ and 211α lattice planes. The phase transformation from autenite to martensite is clearly indicated by the appearance of martensite diffraction lines at a certain temperature. This temperature was identified as $M_s = 180°C$. Additional, a slight movement of the diffraction lines to higher energies is observable during cooling due to decreasing thermal lattice spacing.

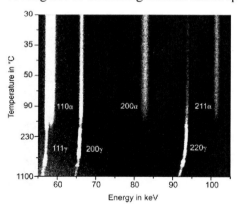

Fig. 2. 2D-density plot during cooling of the LTT alloy. Martensite transformation is indicated by appearing α diffraction lines.

After complete cooling down to ambient temperature (30°C) the austenite diffraction lines are still visible indicating the presence of retained austenite since M_f was not reached and is located below ambient temperature. Once the M_s-temperature is known the content of retained austenite can be estimated by using the Koistinen and Marburger equation [14]:

$$V_\gamma = \exp[-1.10 \cdot 10^{-2}(M_s - T_q)] \qquad (2)$$

V_γ – content of retained austenite
M_s – martensite start temperature
T_q – quenching temperature

Following this equation by applying M_s = 180°C and T_q = 30°C as quenching temperature the retained austenite content should be around 19 %. That issue was validated by determination of the content of retained austenite by using eq. 1 applying the diffraction data. The result is shown in fig.3 incorporating the corresponding temperatures. The final content of retained austenite shows a value of 14 % ± 5 % what is in good correlation with the estimated value of 19 %. Furthermore the rate of martensite transformation follows the slope predicted by the Koistinen-Marburger equation in good agreement. Hence, this equation was supposed to be suitable for describing the martensite transformation behavior of the LTT alloy in quality and quantity. The kinetic of the martensite transformation is characterized by an intense increase immediately after M_s was reached. The material shows already 50 % of martensite at a temperature 60°C below M_s. Consequently the remaining austenite will transform only at slow transformation rates at considerably low temperatures.

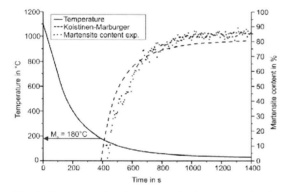

Fig. 3. Transformation from austenite into martensite during cooling cycle. Martensite content measured compared to estimation by Koistinen and Marburger.

Online Stress Measurement

The reaction forces measured under restraint in the CTW test during welding of the LTT alloy and also the reference filler are represented in fig. 4. For clarity only the cooling cycle from 1100°C to ambient temperature is shown.

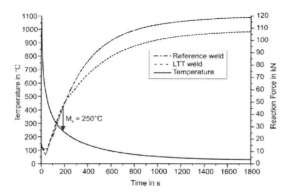

Fig. 4. Reaction force development during cooling of LTT and reference weld from 1100°C down to ambient temperature.

After finishing the welding process the materials begin to cool resulting in a force raise due to shrinkage restraint of the surrounding cooler areas. The increase of the reaction force is higher at this first stage of cooling due to the high temperature gradient. With proceeding cooling the reaction force increase is slowed down ending up with high tensile stresses above 100 kN. In case of the LTT weld the force development is additionally characterized by a reduction of the slope of the reaction force in a certain temperature range leading to an intersection of the former parallel run of both curves. In comparison the final reaction force of the LTT weld is approximately 10% lower as the reference weld. Taking the point in time of the reduction of the reaction force as indication for the beginning of the martensite transformation the M_s-temperature can be identified to be around 250°C. This is 70°C higher than the M_s-temperature determined during diffraction experiments. Chemical dilution with the base material should be one reason for the increased value. However the chemical composition of the weld measured by spark source spectroscopy showed values very close to the composition of the undiluted alloy (table 6). Another reason could be the fact that the impact of the phase transformation is detectable not before some distinct amount of martensite has formed. However similar M_s-temperatures where found during in-situ diffraction experiments during real welding process [15].

Table 6. Chemical composition of LTT weld metal incorporating dilution

C	Cr	Ni	Si	Mn	Fe
0.07	10.1	7.6	0.4	1.0	balance

Applying again the Koistinen-Marburger equation using the M_s-temperature indicated by the reaction force the martensite transformation kinetic was compared to the reaction force reduction during cooling. For that purpose the difference of the reaction forces of the LTT weld and the reference weld was determined. This difference is compared to the martensite formation after Koistinen and Marburger in fig. 5. It becomes clear that the force reduction clearly correlates to the amount of martensite formed during cooling. The retained austenite content at ambient temperature shows around 9 %.

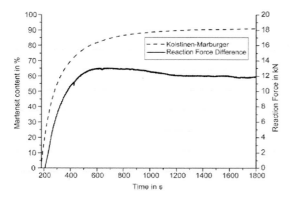

Fig. 5. Detailed view on the difference in reaction forces between LTT weld and to high strength reference weld compared to martensite content determined by Koistinen-Marburger equation.

The results indicate that the martensitic transformation at considerably low temperature has a pronounced impact on the global specimen stressing during and after welding. The rate of stress reduction can be correlated directly to the martensite transformation rate. The results of the metallographic analysis are given in fig. 6 and fig. 7.

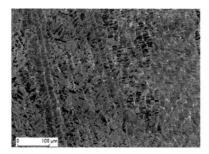

Fig. 6. Etchant after Lichtenegger and Bloech (LBI) revealing predominantly martensitic microstructure

The microstructure after welding consists of martensite and retained austenite. The cell cores are predominantly martensitic. The austenite is present network-like within intercellular areas.

Fig. 7. Microstructure showing intercellular austenite in last solidifying areas between martensitic cells

The primary solidification mode is austenitic and cellular. The intercellular areas solidify at last containing higher chromium and nickel contents compared to the cell cores what was proven by EMPA analysis. These high amounts of alloying elements suppress the martensite formation even at ambient temperature leading to amounts of retained austenite up to 19 % found in this study.

Summary

Experiments were conducted using a novel welding alloy of a characteristic chemical composition leading to martensite transformation in a considerably low temperature range. This Low Transformation Temperature (LTT) alloy was characterized concerning martensite transformation kinetic during a furnace experiment using high energy synchrotron diffraction. For the undiluted alloy the M_s-temperature was located at approximately 180°C. The martensite transformation was not found to be finished at ambient temperature resulting in amounts of retained austenite of approximately 14 %. This is in good agreement to the estimation following Koistinen and Marburger. The rate of martensite transformation is characterized by an exponential decrease. The impact of the phase transformation on the specimen stressing during welding and subsequent cooling was analyzed using a special welding test setup allowing for online observation of the reaction forces. Results showed that the martensite transformation at considerably low temperature leads to a decrease of the final reaction force compared to conventional high strength steel weld by around 10 %. However the superposition of the phase transformation impact on the reaction force build-up caused by thermal shrinkage could be accurately detected. In good correlation with the findings from the furnace experiments the reaction force reduction by the martensite transformation clearly follows the martensite formation rate. That means most of the stress reduction is achieved during the very first stage of transformation. Examinations of the microstructure revealed that the retained austenite is located in intercellular areas resulting from austenitic solidification mode.

Acknowledgments The author would like to especially acknowledge Prof. Christoph Genzel, Dr. Manuela Klaus, Dr. Jens Gibmeier, Dr. Thomas Kannengiesser and Ms. Mohammadzadeh for their support. The author is also grateful to the German Research Foundation DFG for financial funding.

References

[1] Wohlfahrt H (1986) Die Bedeutung der Austenitumwandlung für die Eigenspannungsentstehung beim Schweißen. Härterei Technische Mitt 41:248-257

[2] Jones W K C, Alberry P J (1977) A model for stress accumulation in steels during welding. In: Residual stresses in welded construction and their effects. The Welding Institute, Cambridge

[3] Bühler H, Scheil E (1933) Zusammenwirken von Wärme- und Umwandlungsspannungen in abgeschreckten Stählen. Archiv des Eisenhüttenwesens 6:283-288

[4] Bhadeshia H K D H (2004) Developments in Martensitic and Bainitic Steels: Role of the Shape Deformation. Mater Science and Engineering A 378:34-39

[5] Murata H, Kato N, Tamura H (1993) Proportion of Transformation Superplasticity and Expansion on Stress Releasement. Transactions of JWRI 24:119-124

[6] Ohta A, Watanabe O, Matsuoka K et al (1999) Fatigue Strength improvement by using newly developed low transformation temperature welding material. Weld in the World 43:38-42

[7] Ohta A, Maeda Y, Suzuki N (2001) Fatigue Life Extension by Repairing Fatigue Cracks Initi-ated around Box Welds with Low Transformation Temperature Welding Wire. Weld in the World 45:3-8

[8] Shiga C, Yasuda H Y, Hiraoka K et al (2010) Effect of Ms temperature on residual stress in welded joints of high- strength steels. Weld in the World 54:71-79

[9] Martinez Diez F (2008) Development of Compressive Residual Stress in Structural Steel Weld Toes by Means of Weld Metal Phase Transformations. Weld in the World 52:63-78

[10] Francis J A, Kundu S, Bhadeshia H K D H et al (2009) The Effects of Filler Metal Transformation Temperature on Residual Stresses in a High Strength Steel Weld Intervening Transformations. J of Pres Vessel Technology 131:1-15

[11] Kannengiesser T, Kromm A, Rethmeier M, et al (2009) Residual Stresses and In-situ Measurement of Phase Transformation in Low Transformation Temperature (LTT) Welding Materials. Advances in X-ray Analysis 52:755-762

[12] Genzel C, Denks I A, Gibmeier J (2007) The materials science synchrotron beamline EDDI for energy-dispersive diffraction analysis. Nucl Instrum Methods in Phys Research A 578:23-33

[13] Laine E (1978) A high-speed determination of the volume fraction of ferrite in austenitic stain-less steel by EDXRD. J Phys F Metal Phys 8:1343-1348

[14] Koistinen D P, Marburger R E (1959) A general equation prescribing the extent of the austenite-martensite transformation in pure iron-carbon alloys and plain carbon steels. Acta Metallurgica 7:59-60

[15] Kromm A, Kannengiesser T, Gibmeier J (2010) In-situ Observation of Phase Transformations during Welding of Low Transformation Temperature Filler Material. Mater Science Forum 638-642:3769-3774

Degradation Mechanism of Creep Strength Enhanced Ferritic Steels for Power Plants

K. Sawada, M. Tabuchi and K. Kimura

National Institute for Materials Science, Japan

Abstract

Creep strength degradation of Creep Strength Enhanced Ferritic (CSEF) steels was investigated, focusing on microstructural degradation. The creep strength of ASME Gr.91 steel remarkably decreased in the long-term at 600°C and 650°C. If the data in the long-term are selected for regression analysis, we can accurately evaluate long-term creep strength. The data under applied stresses larger than proportional limit stress should be omitted for the analysis. Creep deformation under a stress lower than proportional limit stress is strongly affected by microstructural changes due to diffusion. After long-term creep exposure at 600°C, the martensitic lath structure was collapsed by recovery and coarsening of precipitates occurred. The recovery of lath structure preferentially occurred around prior austenite grain boundaries. The Z-phase particles nucleated around prior austenite grain boundaries during creep exposure, consuming fine MX particles that were main strengthener.

Introduction

Creep Strength Enhanced Ferritic (CSEF) steels are widely used for components of thermal power plants because of their higher creep resistance. Ultra Super Critical (USC) power plant with higher thermal efficiency was realized by means of the development of CSEF steels [1]. Furthermore, the CSEF steel is candidate as a structural material for the next generation fast reactor and fusion reactor in Japan [2,3]. However, the allowable tensile stresses of CSEF steels recently have been reduced in Japan [4,5], since it became clear that the creep strength of CSEF steel abruptly dropped in the long-term [6]. Therefore, for construction of new power plant, the components should be designed based on the revised allowable tensile stress. In case of operation of present power plants, the component can be used for long time (more than 10 years) based on residual life assessment. Degradation mechanism during long-term creep exposure should be clarified to assess residual life of the component. The design life of the next generation fast reactor is considered to be 60 years in Japan [2]. For the design, we have to predict the creep strength for 60 years because the creep test for 60 years is not realistic. We have to know how the creep strength degrades in the long-term based on degradation mechanism, since the creep strength abruptly drops in the long-term as mentioned above.

The CSEF steel has a tempered martensite stabilized by $M_{23}C_6$ carbide and MX carbonitrides. Creep deformation causes the recovery of martensitic lath and coarsening of the precipitates. These microstructural changes indicate the decrease in deformation resistance, since the martensitic lath boundaries and the precipitates are obstacle to dislocation motion during creep [7]. In addition, the Laves phase and Z-phase are formed during creep exposure. The Z-phase is harmful phase to creep strength [8,9].

This paper clarifies the feature of microstructural changes after long-term creep exposure, focusing on the recovery of martensitic lath structure and precipitates distributions.

Experimental procedure

The material examined was ASME Gr.91 steel. The chemical composition and heat treatment condition are listed in Table 1. The steel has a tempered martensite structure with high dislocation densities after heat treatment. Creep tests were per-

Table 1: Chemical composition (mass%) and heat treatment condition of the steel examined.

C	Si	Mn	P	S	Ni	Cr
0.0900	0.29	0.35	0.009	0.002	0.28	8.70
Mo	Cu	V	Nb	Al	N	
0.90	0.032	0.22	0.072	0.001	0.044	
	Normalizing			Tempering		
	1050°C, 10min A.C.			765°C, 30min A.C.		

formed under a constant load in air, using the specimens with 30mm in gauge length and 6mm in gauge diameter. The specimens for the creep tests were cut longitudinally from center of the wall thickness for Gr.91 tube. The hardness was measured in the gauge portion of crept samples. The specimens for microstructural observation were longitudinally cut from the crept samples in a direction parallel to the stress axis. The microstructure of thin foil and carbon extracted replica was observed, using TEM and STEM-EDX.

Results and Discussion

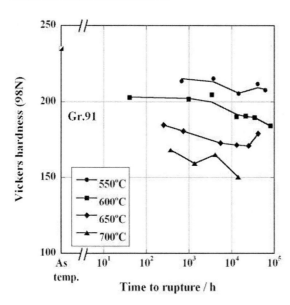

Fig.1: Creep rupture strength of Gr.91 steel.

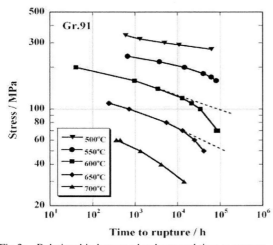

Fig.2: Relationship between hardness and time to rupture.

Creep rupture strength

Figure 1 shows the creep rupture strength of Gr.91 steel at 500°C to 700°C. At 500°C and 550°C, the creep life (time to rupture) was monotonically increasing with decreasing applied stress. On the other hand, the creep life in the long-term was shorter than that (dashed line) predicted from the creep life in the short-term at 600°C and 650°C. This indicates that the creep strength dropped in the long-term at 600°C and 650°C. For accurate prediction of the long-term creep strength, the degradation mechanism of creep strength should be clarified in terms of microstructural changes. Figure 2 shows the relationship between the hardness of gauge portion and the time to rupture. The hardness decreased after creep exposure. The decrease was larger for longer time corresponding to lower stresses. The microstructural changes can cause the decrease in hardness. It can be expected by the hardness result that the microstructural changes are more obvious in the long-term.

Change in dislocation structure during creep exposure

The microstructure of Gr.91 steel consists of a fine lath and a high dislocation density before creep shown in Fig.3. The mean lath width and dislocation density inside lath grain was 0.34 μm and 6.1×10^{14} m^{-2}, respectively. Figure 4 shows the martensitic lath structure after creep exposure at 600°C together with initial lath structure. In the case of sample ruptured under 160MPa (short-term), the lath widths increased after creep exposure. The recovery of lath structure occurred homogenously. On the other hand, large recovery of lath structure was

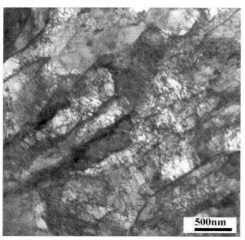

Fig.3: Dislocation structure before creep.

Fig.4: Change in dislocation structure after creep exposure.
   (a) before creep,
   (b) after creep (600oC / 160MPa, tr=971.2h)
   (c) after creep (600oC / 100MPa, tr=34141.0h) T
   he white line indicates prior austenite grain boundary.

observed around prior austenite grain boundary (PAGB) in the case of 100MPa (long-term). This causes decrease in deformation resistance around PAGB, since the lath boundary can be obstacle to dislocation motion during creep [7]. The stress (160MPa) cor responding to high stress condition in Fig.4 is larger than proportional limit (150 MPa) at 600°C. Therefore, the plastic deformation can easily occur without assist of diffusion in case of high stress condition. The stress for low stress condition in Fig.4 is lower than the proportional limit, indicating that the assist of diffusion is needed to cause creep deformation. It can be predicted that the lath structure easily recovers around PAGB since grain boundary diffusion is normally faster than lattice diffusion.

Consequently, the large recovery around PAGB can be a reason for the premature failure in the long-term shown in Fig.1.

Change in precipitates during creep exposure

Figure 5 demonstrates the precipitates distribution after heat treatment using carbon extraction replica. In bright field image, a large number of fine precipitates

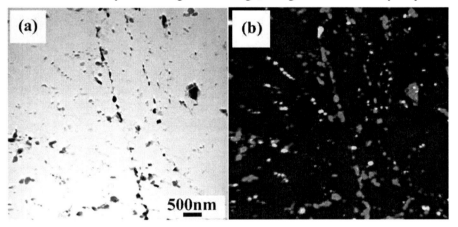

Fig.5: Precipitate distribution before creep.
(a) bright field image, (b) color map based on elemental maps

were observed along PAGB, lath and block boundaries. The result of elemental map is also shown in Fig.5. The map was obtained from superimposing Cr, V and Nb maps. The Cr, V and Nb maps indicate the distributions of $M_{23}C_6$ [M:Cr,Fe,Mo], V-rich MX [M:V] and Nb-rich MX [M:Nb], respectively. The particles shown in red, blue and green correspond to $M_{23}C_6$, V-rich MX and Nb-rich MX in Fig.5. The sizes of MX particles are finer than those of $M_{23}C_6$ particles. Most of particles are mainly distributed along PAGBs, lath and block boundaries, indicating that the particles pin the boundaries. The pinning effect contributes to the strengthening of Gr.91 steel. The precipitates distributions after creep exposure at 600°C are shown in Fig.6. In bright field image, the sizes of particles are larger than those of particles before creep (Fig.5), meaning coarsening of precipitates during creep exposure. The pink particles are shown in the elemental map. The particles are called Z-phase [C(V,Nb)N]. The Z-phase can be formed consuming MX particles during aging or creep exposure because of similarity of their chemical compositions [8,9]. It has been reported that the chemical composition of MX

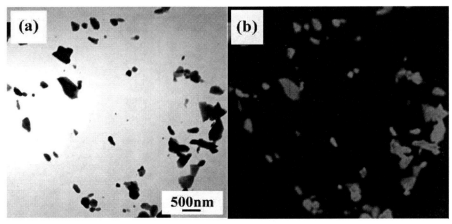

Fig.6: Precipitates distribution after long-term creep.(600₀C / 70MPa, tr=80736.8h) (a) bright field image, (b) color map based on elemental maps

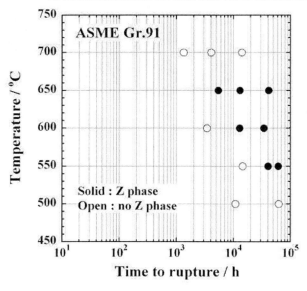

Fig.7: Time-Temperature-Precipitation diagram of Z-phase nucleation.

particle gradually changed, and finally became that of Z-phase during aging [10]. Most of MX particles (blue particle in Fig.5) are disappeared in Fig.6 since a large number of Z-phase particles are formed. The sizes of Z-phase particles are almost the same as that of $M_{23}C_6$ particle in Fig.6. The number density of Z-phase particles can be low due to the large size of Z-phase. Therefore, the strengthening due to Z-phase formation is not large compared with the MX particle. The formation of Z-phase decreases the strengthening effect of MX particles, meaning that the Z-phase is harmful to creep resistance. Figure 7 shows time-temperature-precipitation (TTP) diagram of Z-phase nucleation at several temperatures. The nose temperature of Z-phase nucleation was 650°C. The time required for Z-phase nucleation was about 10000h at 600°C. For the discussion on degradation mechanism in the long-term, Z-phase formation should be considered because the formation became obvious in the long-term.

Relationship between microstructural changes and creep strength

As shown in Fig.8, the stress versus time to rupture curve can be divided into two regions by the proportional limit [2,3]. By means of this region splitting, we can evaluate accurate creep strength in the long-term. As mentioned in section 3.2, the martensitic lath structure homogeneously recovered under the high stress. The preferential recovery around PAGB occurred under the low stress. In short, the degradation mechanism is different between the high stress and low stress region. The critical stress is proportional limit at each temperature. The Z-phase formation occurred after long time, indicating that degradation process under a stress lower than proportional limit can be affected by the Z-phase formation. In addition to the preferential recovery around PAGB, the Z-phase formation together with the disappearance of MX particles contributes to the creep strength degradation in the long-term.

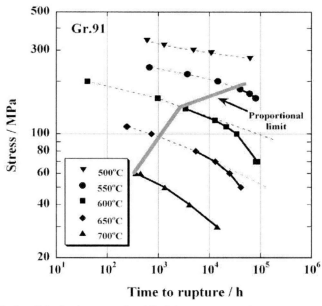

Fig.8: Criterion by proportional limit in stress versus time to rupture curve.

Conclusions

The mechanism for creep strength degradation of ASME Gr.91 steel was investigated, focusing on the recovery of martensitic lath structure and precipitates distributions. The results are summarized as follows.

At 600°C and 650°C, the creep life in the long-term (low stress) was shorter than that extrapolated from the trends of short-term (high stress) creep strength, meaning that the premature failure occurred.

At 600°C, the martensitic lath homogeneously recovered during creep under the high stress (short-term). On the other hand, the lath preferentially recovered around prior austenite grain boundaries during creep under the low stress condition (long-term). The critical stress was the proportional limit at 600°C.

The Z-phase particles were formed after long-term creep together with disappearance of MX particles. The time required for the Z-phase nucleation was 4000h and 10,000h at 650°C and 600°C, respectively.

The stress versus time to rupture curve can be divided into two regions by means of the proportional limit at each temperature, since the degradation mechanism was different between the two regions. The creep strength in the long-term can be accurately evaluated by this region splitting.

References

[1] Masuyama F. History of Power Plants and Progress in Heat Resistant Steels. ISIJ Int., 41, 612-625 (2001)

[2] Onizawa T, Wakai T, Ando M and Aoto K: Proc. Creep and Fracture in High Temperature Components—Design and Life Assessment Issues, ed. by I. A. Shibli *et al.*, DEStech Publications, Inc., Lancaster, USA, (2005), 130.

[3] Tanigawa H et al. Technical issues of reduced activation ferritic/martensitic steels for fabrication of ITER test blanket modules. Fusion Engineering and Design 83, 1471-1476 (2008)

[4] Kimura K. Assessment of long-term creep strength and review of allowable stress of high Cr ferritic creep resistant steels. Proc. PVP2005, 2005 ASME Pressure Vessels and Piping Division Conference, Denver, Colorado USA; July 17-21 2005.

[5] Kimura K. Creep strength assessment and review of allowable tensile stress of creep strength enhanced ferritic steels in Japan. Proc. PVP2006/ICPVT-11, 2006 ASME Pressure Vessels and Piping Division Conference, Vancouver, BC, Canada; July23-27 2006.

[6] Kushima H, Kimura K, Abe F. Degradation of Mod.9Cr– 1Mo Steel during long-term creep deformation. Tetsu-to-Hagane 85, 841-847 (1999)

[7] Maruyama K, Sawada K and Koike J. Strengthening mechanisms of creep resistant tempered martensitic steel. *ISIJ Int.*, **41** 641-653 (2001)

[8] Hald J. Metallography and alloy design in the COST 536 action. In: Lecomte-Beckers J, et al, editor. 8th Liege Conf. on Materials for Advanced Power Engineering. Jülich:Forschungszentrum, Jülich GmbH; 2006.

[9] Sawada K, Kushima H, Kimura K, Tabuchi M. TTP diagrams of Z phase in 9-12%Cr heat resistant steels. ISIJ Int 47, 733-739 (2007).

[10] Cipolla L, Danielsen H.K, Venditti D, Nunzio P.E.Di, Hald J and Somers M.A.J. Conversion of MX nitrides to Z-phase in a martensitic 12% Cr steel. Acta Mater., 58, 669-679 (2010)

Electrochemical studies on pitting corrosion on Cr13 steel exposed to CO_2 and artificial brine with high chloride concentration

Oleksandra Yevtushenko, Ralph Bäßler

BAM Federal Institute for Materials Research and Testing, Unter den Eichen 87, 12205 Berlin, Germany

Abstract

Within the project COORAL (German acronym for "CO_2 purity for capture and storage") first studies on piping steels exposed to CO_2 and artificial brine with high chloride concentration have been carried out. Corrosion behavior of martensitic Cr13 steel (1.4034) was investigated in a corrosive environment (artificial saline brine, $T = 60\,°C$, CO_2-flow rate $3-5$ L/h, atmospheric pressure, exposure times from 1 h up to 14 days) using electrochemical and metallographic techniques. Different corrosion kinetics were observed as a function of exposure times and chloride concentration in the artificial brine. In CO_2-saturated brine pitting corrosion was observed at free corrosion potential, whereas in the brine without addition of NaCl a stable passive layer built up. Predictions about corrosion mechanism are made and verified by means of surface analytical techniques.

Keywords: stainless steel, pitting corrosion, electrochemistry, carbon dioxide

Introduction

Carbon dioxide induced corrosion on the piping steels can cause problems during injection of emission gasses from combustion processes in deep geological layers (Carbon Capture and Storage, CCS). In severe operation conditions, containing CO_2, corrosion resistant alloys, like Cr13-steel, are commonly used [1]. However, in deep geological thermal fluids, these steels can be ineffective since the chloride concentration and temperature could be extremely high. The aquifer fluid saturated with CO_2 and with high chloride concentration can cause the pitting corrosion on the piping steels [2, 3]. Therefore, material selection of appropriate piping steels is a key factor in order to increase the safety and the reliability of the CCS-technology, and for the reduction of the process costs.

The aim of this contribution is to investigate the corrosion behaviour of 13 % Cr-steel in CO_2 saturated artificial brine with different chloride concentrations known from Stuttgart Aquifer in terms of electrochemical exposure experiments (up to 7 days).

Materials and methods

The specimens (20 x 8 x 3 mm), prepared from the martensitic thermally treated steel X46Cr13 (1.4034, AISI 420), were grinded with 320 SiC-paper, thoroughly cleaned with deionised water and rinsed with acetone in order to remove residual oil from the surface. A new specimen was used for every experiment. The experiments were carried out in artificial aquifer brine with the following chemical composition: 1760 mg/L Ca^{2+}, 430 mg/L K^+, 1270 mg/L Mg^{2+}, 90100 mg/L Na^+, 143300 mg/L Cl^-, 3600 mg/L SO_4^{2-}, 40 mg/L HCO_3^- dissolved in bi-distilled water, pH_{brine}= 8.9. Before each experiment the brine was heated up to 60 °C and saturated with technical CO_2, which was bubbled through the solution for minimum 1 h before each experiment until the saturation with CO_2 is achieved, and the pH of 6 ± 0.1 did not change its value. For the investigation of the role of chloride ions in the process of pitting corrosion the electrolyte was prepared without addition of NaCl. After the saturation with technical CO_2 the pH was measured to be 5.8. The CO_2 flow rate of 3 - 5 L/h was controlled by a commercial flow meter (VAF Fluid Technik GmbH).

Electrochemical measurements were carried out using Gamry Potentiostatic System Model Reference 600 in a standard three electrodes cell. Ti/TiO_2 net counter electrode and a saturated Ag/AgCl (E = 0.3 V vs. SHE) reference electrode were used. The Electrochemical Impedance Spectroscopy (EIS) measurements were carried out using 10 mV amplitude perturbations and a frequency range of 10 kHz to 10 mHz at the open circuit potential measured for a minimum of 1 h before every experiment to insure the steady state value.

Results and discussion

It has been reported that passive film of stainless steels with high chromium content is mainly a chromium oxide, which protect the steel from corrosion attack in an aggressive environment [4, 5]. Chromium has high metal-oxygen bond strength which is necessary for a stable passive film formation.

In order to investigate the corrosion mechanism of the steel X46Cr13 the series of EIS measurements has been carried out at the open circuit potential at different exposure times. In Fig. 1 the Bode-plots and Nyquist-diagrams of the specimen obtained at different times during 24 h immersion in the artificial brine are presented. To obtain the complex information on a corrosion system an adequate equivalent circuit, which represents the impedance spectrum of the present corrosion system, must be found. In Fig. 2 the simple equivalent circuit for corroding metal is shown, where R_{sol} describes the electrolyte resistance, R_{ct} – charge transfer resistance, Y and a represent the constant phase element, which is widely used in data fitting for depressed semicircles. The physical origin of depressed semicircles may lie on the surface imperfections which cause not ideal capacitive behaviour of the corroding metal surface [6]. In Fig. 1 the solid lines represent the fits by using

Electrochemical studies on pitting corrosion on Cr13 steel exposed to CO_2

of the above mentioned equivalent model, the fitting results are presented in Table 1. The increase of the corrosion rate with the exposure time can clearly be

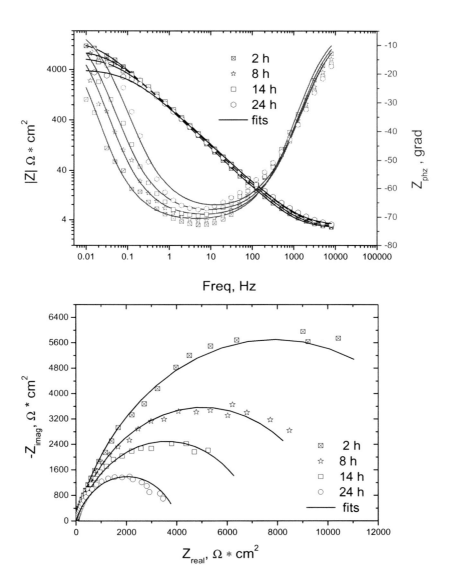

Fig. 1: Bode- and Nyquist-plots of the X46Cr13-steel during the 24 h exposure to CO_2-saturated saline brine. The symbols represent experimental data, the solid lines represent the fits by the model presented in Fig. 2.

Table 1: Values obtained by fitting the impedance data measured on the specimen exposed for 24 h to CO_2-saturated saline brine

Parameter	2 h Value	2 h Error	8 h Value	8 h Error	14 h Value	14 h Error	24 h Value	24 h Error	Units
R_{ct}	15.8E+03	68.07	10.1E+03	33.6	7.20E+03	29.92	4.08E+03	13.63	$\Omega\,cm^2$
R_{sol}	2.69E+00	6.79E-03	2.76E+00	7.35E-03	2.88E+00	8.11E-03	3.01E+00	8.58E-03	$\Omega\,cm^2$
Y	3.00E-04	8.79E-06	2.50E-04	9.10E-06	3.00E-04	1.02E-05	3.25E-04	1.21E-05	$S*s^a/cm^2$
a	7.98E-01	1.40E-03	7.86E-01	1.47E-03	7.72E-01	1.62E-03	7.62E-01	1.77E-03	
C_{true}	448.8E-06		321E-06		376E-06		355E-0.6		F/cm^2

seen: the R_{ct} value changes from 15.8 kΩ cm^2 down to 4.08 kΩ cm^2 during the 24 h of the immersion time indicating an active corroding surface, as evidence several pits were observed on the specimen surface at that time (see Fig. 3).

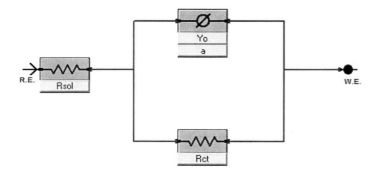

Fig. 2: Equivalent circuit model with one time constant

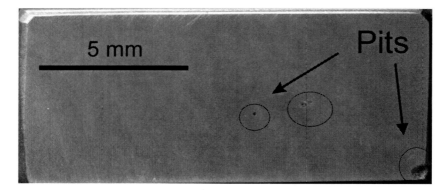

Fig. 3: X46Cr13-steel specimen after 24 h exposure in CO_2-saturated saline brine

It is known that spontaneous pitting of passive metals usually starts at sharp edges, burrs, and in the places, where the passivating oxide film is thin or otherwise defect [7]. At the same time it also can be caused by aggressive surrounding environment like, for example, the chloride ions or the carbonic acid formed immediately by the contact of CO_2 with the water. In CO_2-saturated solutions the corrosion rate is expected to decrease with the exposure time due to the formation of protective carbonate scales on the steel surface [8, 9, 10]. However, if Ca^{2+} present in the solution the early precipitation of $CaCO_3$ can prevent or alter the subsequent precipitation of protective $FeCO_3$ [11]. According to the X-ray diffraction analysis of the X46Cr13-steel surface, after 24 h of exposure to CO_2-saturated saline brine, no evidence of the $FeCO_3$-phase was found on the surface.

The role of chloride ions in the pitting corrosion has been widely discussed in the literature [12] as well as the pitting corrosion mechanism in CO_2-saturated saline solutions [13, 14]. If Cl^--ions are present in the electrolyte they compete with OH^- for adsorption on surface sites. If Cl^- are adsorbed at the oxide surface, the metal-chloride or metal-oxide-hydroxide-chloride surface complexes are less strongly bonded to the oxide matrix, and the activation energy for their transfer to

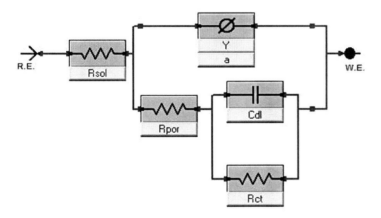

Fig. 4: Equivalent circuit model with two time constants

the electrolyte is decreased. As a result the localized dissolution rate is increased and pits start to grow. After local depassivation, the repassivation will be poisoned by the competitive adsorption mechanism between chloride and hydroxide. Therefore, it has been suggested that the origin of nucleated pits on the X46Cr13-steel in the investigated artificial brine is due to the high chloride concentration. As a main supplier of chloride ions the NaCl was excluded from the electrolyte's preparation procedure. The artificial brine without NaCl was heated up to 60 °C and saturated with CO_2. after the saturation the pH was measured to be 5.8. In Fig. 5 the Bode-plots and Nyquist-diagrams of the X46Cr13–specimen obtained during 7 days immersion in the artificial brine without NaCl are presented. The solid lines represent the fits due to the model shown in Fig. 4, where R_{sol}, R_{por}, R_{ct} describe

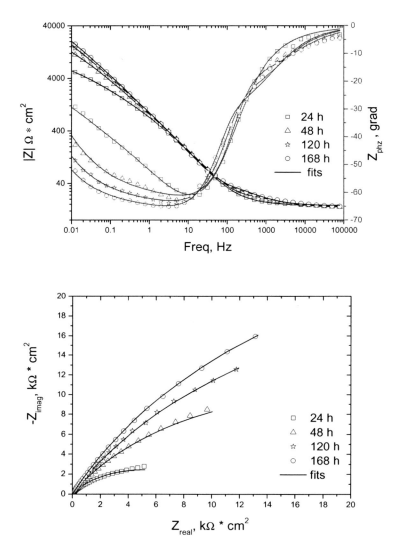

Fig. 5: Bode- and Nyquist-plots of the X46Cr13-steel during the 168 h exposure to CO_2-saturated saline brine without NaCl. The symbols represent experimental data; the solid lines represent the fits by the model presented in Fig. 4.

the electrolyte solution resistance, the resistance of the passive layer and charge transfer resistance through the electrode/layer interface, respectively. The C_{dl} represents the double layer capacitance, and the constant phase element describes the frequency dependent effects occurring on the electrolyte/passive layer interface. The appearance of the second time constant in the equivalent circuit can be explained by the formation of the porous corrosion layer, probably consisting on the iron/chromium oxides/hydroxides. The fitting results are presented in Table 2. The increase of the semicircles with the exposure time

Table 2: Values obtained by fitting the impedance data measured on the specimen exposed for 168 h to CO_2-saturated brine without NaCl

Parameter	24 h		48 h		Units
	Value	Error	Value	Error	
R_{por}	6.80E+00	1.08E-01	2.19E+01	3.69E-01	Ω cm^2
R_{ct}	1.04E+04	1.09E+02	3.48E+04	5.03E+02	Ω cm^2
R_{sol}	1.32E+01	4.52E-02	1.45E+01	3.68E-02	Ω cm^2
Y	4.17E-04	2.04E-05	3.13E-04	1.19E-05	S*s^a/cm^2
a	5.46E-01	6.04E-03	6.33E-01	4.41E-03	
$C_{'true'}$	3.18E-06		4.58E-06		F/cm^2
C_{dl}	3.90E-05	4.77E-06	2.50E-05	5.24E-06	F/cm^2
Parameter	120 h		168 h		Units
	Value	Error	Value	Error	
R_{por}	3.99E+01	7.60E-01	5.42E+01	9.71E-01	Ω cm^2
R_{ct}	6.41E+04	1.25E+03	10.1E+04	2.62E+03	Ω cm^2
R_{sol}	1.48E+01	3.55E-02	1.56E+01	3.46E-02	Ω cm^2
Y	2.70E-04	1.09E-05	2.51E-04	1.07E-05	S*s^a/cm^2
a	6.62E-01	3.73E-03	6.79E-01	3.32E-03	
$C_{'true'}$	8.24E-06		11.0E-06		F/cm^2
C_{dl}	2.50E-05	5.50E-06	2.90E-05	5.71E-06	F/cm^2

can be clearly seen in Nyquist-plots indicating the decrease of the corrosion rate. The value of R_{ct} increases from 10 kΩ cm^2 after 24 h exposure time up to 100 kΩ cm^2 after 168 h, the value of R_{por} increases from 6.8 Ω cm^2 at 24 h up to 54.2 Ω cm^2 after 168 h. At the same time the value of the "true" capacitance increases from 3 μF/cm^2 up to 11 μF/cm^2. These observations lead to the conclusion that in CO_2-saturated brine without addition of NaCl a passive layer is formed, consisting probably of inner dense chromium/iron oxide layer and outer chromium porous hydroxide layer, which grows and compresses in expense of the inner pores or gaps during the exposure time. Due to the AES surface analysis thin chromium-rich oxide film was found on the specimen´s surface.

Conclusion

Initial stages of corrosion process occurring at the X46Cr13-steel in CO_2-saturated saline brine with different chloride concentration are investigated by means of electrochemical impedance experiments. It has been shown that in CO_2-saturated saline brine corrosion process starts as pitting corrosion induced by the high concentration of chloride ions within the first 24 h of immersion time. In the absence of NaCl a stable passive layer is grown. It must be concluded, that massive care should be taken when putting this steel into operation in environments with high chloride concentration similar to those investigated in the laboratory.

Acknowledgements

This work has been made in the frame of the German research project COORAL ("CO_2 purity for capture and storage"). The authors are grateful for the financial support by German Ministry of Economics.

References

1. Denpo, K. and H. Ogawa, *Fluid flow effects on CO_2 corrosion resistance of oil well materials,*. Corrosion, 1993. **49**(6): p. 442-449.
2. Pfennig, A. and R. Bäßler, *Effect of CO_2 on the stability of steels with 1% and 13% Cr in saline water.* Corrosion Science, 2009. **51**: p. 931-940.
3. Lopez, D.A., T.Perez, and S.N. Simison, *The influence of microstructure and chemical composition of carbon and low alloy steels in CO_2 corrosion. A state –of -the - art appraisal.* Materias & Design, 2003. **24**: p. 561-575.
4. Iversen, A. and B. Leffler, *Aqueous Corrosion of Stainless Steels*, in *Shreir's Corrosion*, M.G. B. Cottis, R. Lindsay, S. Lyon, T. Richardson, D. Scantlebury, H. Stott, Editor. 2010, Elsevier: Amsterdam.
5. Mischler, S., H.J. Mathieu, and D. Landolt, *Investigation of a Passive Film on an Iron-Chromium Alloy by AES and XPS.* Surface and Interface Analysis, 1988. **11**: p. 182-188.
6. Cottis, R. and S. Turgoose, *Electrochemical Impedance and Noise.* Corrosion Testing Made Easy, ed. B.C. Syrett. 1999, Houston: NACE International. 149.
7. Kaesche, H., *Corrosion of Metals: Physicochemical principles and current problems.* Engineering materials and processes, ed. B. Derby. 2003, Berlin Heidelberg: Springer-Verlag
8. Nesic, S., et al., *A mechanistic model for CO_2 corrosion with protective iron carbonate films.* Corrosion2001, 2001. **paper No. 01040**.
9. Linter, B.R. and G.T. Burstein, *Reactions of pipeline steels in carbon dioxide solutions.* Corrosion science, 1998. **41**: p. 117-139.
10. Li, T., et al., *Mechanism of protective film formation during CO_2 corrosion of X65 pipeline steel.* Materials, 2008. **15**(5): p. 702-706.
11. Crolet, J.L. and M.R. Bonis, *Algorithm of protectiveness of corrosion layers. 1- Protectiveness mechanisms and CO_2 corrosion prediction*, in *Corrosion 2010*. 2010, NACE international: San-Antonio.paper No?
12. Maurice, V. and P. Marcus, *Structure, Passivation and Localized Corrosion of Metal Surfaces*, in *Modern Aspects of Electrochemistry*, J.-W.L. Su-Il Pyun, Editor. 2009, Springer: London.
13. Fang, H., B. Brown, and S. Nesic, *High Salt Concentration effect on CO_2 Corrosion and H_2S Corrosion*, in *Corrosion 2010*. 2010, NACE International: San-Antonio.

14. Zhao, G.X., et al., *Effect of Temperature on Anodic Behavior of 13Cr Martensitic Steel In CO_2 Enviroment.* Metals and Materials International, 2005. **11**(2).

Development of $^{10}B_2O_3$ processing for use as a neutron conversion material

L.F. Voss, J. Oiler, A.M. Conway, R.T. Graff, C.E. Reinhardt, Q. Shao, T. F. Wang, and R.J. Nikolic

Lawrence Livermore National Laboratory, Livermore, CA 94550

Abstract

Development of thermal neutron detectors is critical for a number of homeland security and physics applications. In this work, we describe our efforts towards developing boron-10 oxide ($^{10}B_2O_3$) as a thermal neutron conversion material for textured PIN silicon pillar diode platforms. The filling of the textured diode was achieved by first infiltrating the structures with aqueous boric acid, followed by thermal annealing at above the melting point of boron oxide to achieve a high fill and conversion of the materials into boron oxide. The present solution phase processing strategy has the potential to yield amenable and economical manufacturing processes for coating neutron converter materials on detectors. Potential performance of a boron oxide filled detector is also discussed.

Introduction

Interest in thermal neutron detectors has increased significantly over the past several years in response to heightened fears of nuclear proliferation and terrorism. The application space for thermal neutron detectors is broad. Reliable detectors are needed in a large number of form factors with different size and performance requirements. Currently, ^{3}He filled gas tubes are the standard for thermal neutron detection the in field. In response to the diminishing supply of available ^{3}He and the rapidly increasing demand for thermal neutron detectors, a number of technologies are being explored. These include boron-10- (^{10}B) lined gaseous tubes, boron-10-triflouride ($^{10}BF_3$) -filled tubes, lithium-6-(^{6}Li) doped scintillators, and other semiconductor based platforms utilizing ^{6}Li [1-4], ^{10}B or ^{10}B containing compounds [5-10] such as boron nitride, boron carbide [11] and boron phosphide. Our group has been developing a 3-dimenstional (3d) Si based "Pillar Detector," first reported in 2005 by Nikolic et al [8] with ^{10}B as the thermal neutron conversion material. This approach relies on pure ^{10}B deposited by low pressure chemical vapor deposition (LPCVD)[10], and has achieved detectors with 20% intrinsic detection efficiency [9], with the possibility to increase to greater than 50%. McGregor et al. have developed a design based on ^{6}LiF in perforations, also a promising approach based on a 3d geometry.[1-4] Typically, thermal neutron detectors operate on similar principles, relying on an isotope with a high thermal neutron cross-section such as ^{3}He, ^{10}B, ^{6}Li, ^{157}Gd and ^{113}Cd. An incident thermal neutron interacts with the neutron sensitive material, resulting in a nuclear reaction. The products are then either directly collected or converted to electron-hole pairs in a semiconductor, or photons in a scintillator.

The use of a semiconductor based system offers several advantages over gas filled tube based detectors. First, since a solid neutron converter material is used instead of a gas, this allows a significantly smaller detector footprint for solid state detector devices. In addition, a solid detector offers the potential for superior fieldability due to their decreased sensitivity to vibrations, ease of transport compared to gaseous chambers, and potential for lower voltage operation. In the initial development of solid state detectors, the semiconductor based detectors were based on a flat diode coated with a thin film of neutron conversion material. However, due to inherent limitations in this type of design, the maximum efficiency for detectors with a single conversion layer design is about 4%. [1] Recently, our group has reported on a three dimensional solution which has the potential to achieve >50% efficiency [9]. Having a non-vacuum based deposition method is attractive from a process cost-reduction and complexity standpoint. In this work, we report on the use of ^{10}B$_2$O$_3$ as a potential neutron converter material within our 3D pillar structure.

Thermal neutron conversion in boron oxide

Semiconductor based thermal neutron detectors typically go through a three-step detection process (Figure 1a). In the first step, a thermal neutron collides with the neutron conversion material and creates several nuclear or charged particles and gamma rays. In the second step, these charged products deposit their energy in the semiconductor detector in the form of electron-hole pairs. The third step is the collection of these pairs to generate a detectable signal.

In order to determine whether a detector design based on ^{10}B$_2$O$_3$ can feasibly achieve a high intrinsic efficiency , we have calculated the probability of neutron capture for ^{10}B$_2$O$_3$ films of varying thickness. ^{10}B$_2$O$_3$ has a ^{10}B atomic density of only 33.8% that of pure ^{10}B solid, 0.723 g/cm^3 compared to 2.13 g/cm^3. This translates to an increase in neutron mean free path (λ) from 18 μm to 53 μm. To determine the effect of this on the thermal neutron efficiency for homeland security purposes, it must be noted that thermal neutrons to be detected in this case are not strictly directional. This is because neutrons are expected to be emitted at high energies from special nuclear materials and only become thermalized as they are moderated in the surrounding environment. Figure 2a shows that we have opted to calculate efficiency based on a 4π distribution of thermal neutrons, or neutrons incident on the detector equally from all directions and at all angles. Figure 2b dis-

Development of $^{10}B_2O_3$ processing for use as a neutron conversion material

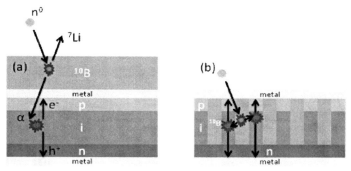

Figure 1: Schematics of (a) 2-D semiconductor neutron detector and (b) 3-D pillar structured neutron detector [6-8].

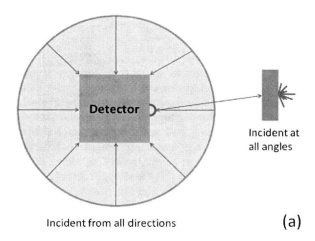

Figure 2: (a) Schematic of simulated neutron environment and (b) neutron capture probability for $^{10}B_2O_3$ films of varying thickness vs. incident angle of neutron.

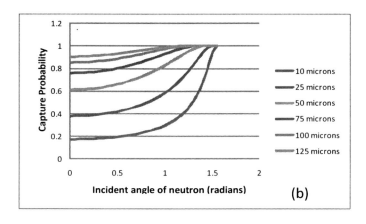

plays the neutron capture probability vs. incident angle for $^{10}B_2O_3$ films of varying thickness. From this, it is clear that a significant fraction of the incident neutrons can be captured only for sufficiently thick coatings. Because our pillar structure relies on interspersed Si p-i-n diodes to overcome the limitation on coating thickness imposed by the Li and alpha travel lengths, we have calculated the path lengths of these particles in $^{10}B_2O_3$ using "Transport of Ions in Matter" (TRIM), shown in Table 1. These are similar to the pillar spacing and pitch (2 μm and 4 μm, respectively) we have previously fabricated and reported for our ^{10}B filled devices. We find that a thickness of >70 μm is needed to capture 85% of the neutrons incident on a film, compared to >95% capture for a 50 μm ^{10}B film.

Table 1: Path lengths of reaction products in $^{10}B_2O_3$

	1.47 MeV Alpha	0.84 MeV Li
$^{10}B_2O_3$	3.5μm	1.89μm

Fabrication of boron oxide filled pillar array

One advantage of semiconductor based detectors is that their development can draw on the extremely strong knowledge developed in the integrated circuit industry. To process our devices, we exploit common semiconductor manufacturing processing tools and techniques including photolithographically defined patterns, dry plasma and wet chemical etching, lapping and polishing of materials, and sputter or e-beam deposition of electrodes. The major challenges to achieving these detectors include etching pillars to sufficient height, achieving a high degree of filling of pillars with $^{10}B_2O_3$, and processing the $^{10}B_2O_3$ after fill in order to deposit electrodes and insulate it from the environment. The last item is particularly important. Since $^{10}B_2O_3$ is hygroscopic, it converts to $H_3^{10}BO_3$ when exposed to humidity and dissolves readily in a large array of polar solvents such as water and methanol.

Silicon pillar array platforms are defined using standard photolithographic techniques. To achieve the aspect ratios required, we employ nLOF 2035 negative photoresist, which produces a mask of ~2.8 μm in thickness. Etching is done in an STS DRIE Bosch system, using a process recipe composed of alternating fluorobutante (C_4F_8) sidewall passivation and sulfur hexafluoride (SF_6) silicon etch steps [8]. Erosion of the photoresist mask is a potentially serious problem if the selectivity is not carefully controlled; an overall etch selectivity of >25:1 must be achieved to produce silicon pillar arrays of greater than 75 μm. In addition, care must be taken not to overestimate the selectivity due to highly aspect-ratio-dependent etching. Also, the sidewall slope of the pillars must be nearly vertical for our device design. For an array consisting of 2-μm wide pillar with a height of >50 μm, even slight positive or negative tapers can result in pinching off of the

Development of $^{10}B_2O_3$ processing for use as a neutron conversion material 59

etch or undercut of the pillar leading to insufficiently tall pillars or collapse of the pillars, respectively. We have achieved >70 μm tall pillars with photoresist remaining on top. Figure 3 shows a cross-sectional scanning electron microscopy (SEM) image of these as-processed pillars.

Figure 3:
SEM image of pillars etched to > 70 μm in height.

Solution based deposition of $^{10}B_2O_3$

Filling such high aspect ratio structures presents a challenge. Common solid source deposition techniques, such as sputtering and e-beam evaporation, generally do not achieve a reasonable fill due to pinching off or shadowing effects at the tops of the pillars. While gaseous chemical vapor deposition is possible and has produced excellent results for our ^{10}B detectors [5-10], material precursors and deposition equipment can be expensive. As an alternative, we have developed processing capabilities to deliver $^{10}B_2O_3$ via a solution-based process to achieve near 100% fill.

$^{10}B_2O_3$ is unique among compounds containing a high boron percentage. It is soluble in a variety of benign polar solvents and is extremely hygroscopic. $^{10}B_2O_3$ kept under normal room conditions will inevitably convert to $H_3^{10}BO_3$. Fortunately, this process can be reversed upon heating, forming several intermediate compounds depending on the temperature, as shown below.

H_3BO_3 -> HBO_2 + H_2O (433K)
$4HBO_2$ -> $H_2B_4O_7$ + H_2O (573K)
$H_2B_4O_7$ -> $2B_2O_3$ + H_2O (Further heating)

In addition, B_2O_3 has a relatively low melting temperature, 450°C, which is low enough to minimize doping of the Si pillars by B diffusion. For comparison, boron implanted in silicon is generally activated using rapid thermal annealing (RTA) at temperatures greater than 1000K. Nonetheless, merely piling on powder and heating the $^{10}B_2O_3$ to 450°C cannot be applied to fill our pillar arrays because it will result in limited infiltration of the pillar structure (Fig 4).

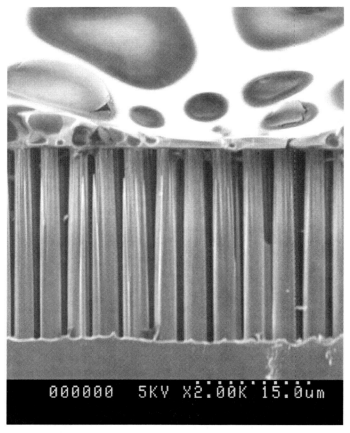

Figure 4: SEM image of B_2O_3 melted onto pillars without aqueous delivery.

Our strategy to achieve a high degree of filling of $^{10}B_2O_3$ in our pillar array structures is to dissolve the $^{10}B_2O_3$ in a solvent in order to infiltrate the pillar structure. In our process, $^{10}B_2O_3$ powder is added to methanol until saturation. The solution mixture is then delivered by pipettes to coat the pillar structured chips. The methanol is allowed to evaporate off entirely under normal room temperature conditions. No extra immediate heat is supplied to the samples because rapid evaporation and/or boiling of the methanol can result in broken pillars. The chip is then inserted into a furnace, subsequently heated to between 450°C and 510°C. After this heat treatment, Raman spectroscopy study is applied to the samples to confirm the presence of $^{10}B_2O_3$ and absence of $H_3^{10}BO_3$ (Figure 5). The Raman peak

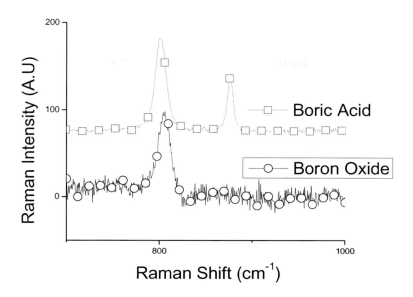

Figure 5: Raman spectra of (a) H$_3$BO$_3$/B$_2$O$_3$ before annealing and (b) B$_2$O$_3$ after annealing.

present at ~805 cm^{-1} indicates B-O bonds [12], while that at ~880 cm^{-1} indicates the O-H bonds indicative of H$_3$BO$_3$ [13]. Note the absence of the latter after heat treatment. Some indication exists that temperatures >500°C lead to reaction of the ^{10}B$_2$O$_3$ with Si, with possible by-products such as SiO$_2$, Si-B compounds, and borosilicates [14]. Though we have observed reductions in pillar size at annealing temperatures larger than 600 °C, no appreciable sample shrinkage effect is found at 500°C, even when held at this temperature for longer than 24 hours. Figure 6a shows a pillar chip of 25-μm height coated with ^{10}B$_2$O$_3$ at 500°C. Figure 6b shows the risk of increasing the annealing temperature. In this case, the sample was heated to 700°C. Note that the pillars have nearly completely dissolved into the surrounding ^{10}B$_2$O$_3$. The "bulbs" on the tops of some pillars are a result of incomplete fill. These areas of silicon have not come into prolonged contact with the surrounding ^{10}B$_2$O$_3$ and so have not reacted. We have also performed filling of ^{10}B$_2$O$_3$ on 50-μm pillar chips with similar success.

Etching of ^{10}B$_2$O$_3$

A serious challenge to the use of ^{10}B$_2$O$_3$ in a detector is developing the capability to planarize the coating for further electrode deposition processing. Our detector requires a blanket metal contact on top of the pillars. The ^{10}B$_2$O$_3$ coating that we deposit results in an uneven over-filled coating on the top of the pillars; the excess material must be removed with a method that does not attack the underlying Si pillars should they become exposed. Both wet and dry etching techniques are unsuitable for this task because neither will return a flat surface. However,

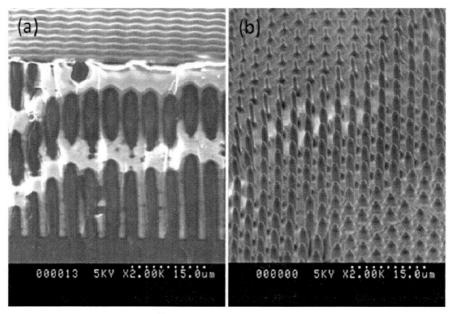

Figure 6: Pillar structure filled with $^{10}B_2O_3$ processed at (a) 500°C and (b) 700°C.

most dry etching recipes that remove $^{10}B_2O_3$ will almost certainly attack the underlying Si. We have opted to develop a process to mechanically etch the $^{10}B_2O_3$. By utilizing a gentle enough polish, the underlying Si should not be damaged. In addition, since higher elevation features are removed faster than lower ones, it should be possible to achieve a final planar final surface for electrode deposition.

Chemical-mechanical planarization (CMP) is a common method for removing overfill of both ductile materials (i.e. metals) [15-16] and brittle materials such as SiO_2 [17]. In addition, it has previously been used to remove borosilicate glass, which generally contains ~13% B_2O_3. However, solutions for etching borosilicate glass are water-based; these etchants rely on hydrolyzing the silicon dioxide for weakening the structure [18], rather than the B_2O_3. When exposed to a water-based solution or slurry, the B_2O_3 begins to dissolve immediately and cracks grow within the oxide fill. Therefore, we have opted to use silicone oil. By using 0.3 μm size grit, a spin speed of 50 rpm, we achieved a $^{10}B_2O_3$ removal rate of ca. 1.5 μm/min. This is a very fast etch rate considering this compound is typically used as a polishing etch which has a neglible etch rate for most group IV and III-V semiconductors. This, however, results in infiltration of the Al_2O_3 grit into the pillar structure, replacing the $^{10}B_2O_3$, shown in Figure 7a. This can be avoided by polishing only with a polishing pad and a small amount of silicone oil. Figure 7b shows a SEM image of a fully etched surface. Electrode deposition and device testing on these boron oxide filled structures are currently in progress.

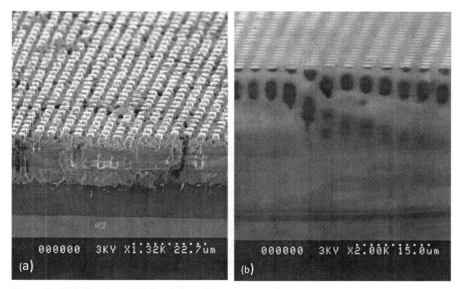

Figure 7: (a) Infiltration of Al$_2$O$_3$ grit into pillar structure and (b) pillar structure filled with ^{10}B$_2$O$_3$ after planarization polishing without grit.

Summary

We demonstrated that boron oxide can be used within our 3d Pillar Platform using a solution based process for the fabrication of thermal neutron detectors. Upon evaluating the composition of ^{10}B$_2$O$_3$, and mean free path of neutrons within this material, capture of a significant number of neutrons can be achieved with pillar height in excess of 70 μm. A simple polishing solution of silicon oil with 0.3 micron "polish grit" is also developed to achieve an etch rate of 1.5 μm /min of ^{10}B$_2$O$_3$ for planarization of our structures after the boron oxide deposition process. Further work on ^{10}B$_2$O$_3$ encapsulation is currently being pursued to minimize the dissolution of boron oxide under humid environment.

ACKNOWLEDGEMENTS
This work was performed under the auspices of the U.S. Department of Energy by Lawrence Livermore National Laboratory under Contract DE-AC52-07NA27344, LLNL-CONF-450383. This work was supported by the Domestic Nuclear Detection Office in the Department of Homeland Security.

References

1. D. S. McGregor, R. T. Klann, H. K. Gersch and Y. H. Yang, Nucl. Instrum. Methods Phys. Res. A **466** (1), 126-141 (2001).
2. D. S. McGregor, R. T. Klann, H. K. Gersch, E. Ariesanti, J. D. Sanders and B. Van Der Elzen, IEEE Trans. Nucl. Sci. **49**, 1999-2004 (2002).

3. J.K. Shultis and D. S. McGregor, IEEE Trans. on Nucl. Sci. **53** (3), 1659-1665 (2006).

4. J.K. Shultis and D.S. McGregor, Nucl. Instrum. Methods Phys. Res. A 606 (3), 608-636 (2009).

5. A. M. Conway, R. J. Nikolic and T. F. Wang, International Semiconductor Device Research Conference, College Park, MD, December 12-14, 2007.

6. R. J. Nikolic, A. M. Conway, C. E. Reinhardt, R. T. Graff, T. F. Wang, N. Deo and C. L. Cheung, International Conference on Solid State and Integrated Circuit Technology, Beijing, China, October 20-23, 2008.

7. R. J. Nikolic, C. L. Cheung, C. E. Reinhardt and T. F. Wang, International Symposium on Integrated Optoelectronic Devices, 6013 (1), 36-44 (2005).

8. R. J. Nikolic, A. M. Conway, C. E. Reinhardt, R. T. Graff, T. F. Wang, N. Deo and C. L. Cheung, Appl. Phys. Lett. 93, 133502 (2008).

9. A. M. Conway, L. F. Voss, C. E. Reinhardt, R. T. Graff, T. F. Wang, N. Deo, C. L. Cheung, and R. J. Nikolic, CMOS Emerging Technologies Conference, February 2009.

10. Deo, N., Brewer, J.R., Reinhardt, C.E., Nikolić, R.J. and Cheung, C.L, J. Vac. Sci. Technol. B, 26, 1309-1314 (2008).

11. K. Osberg, N. Schemm, S. Balkir, J. I. Brand, M. S. Hallbeck, P. A. Dowben and M. W. Hoffman, IEEE Sensors J. **6** (6), 1531-1538 (2006).

12. D. Maniu, T. Iliescu, I. Ardelean, S. Cinta-Pinzaru, N. Tarcea, and W. Kiefer, J. Mol. Struct. 651 (1), 485 (2003).

13. R. R. Servoss and H. M. Clark, J. Chem. Phys. 26 (5), 1175 (1957).

14. E. De Fresart, S. S. Rhee, and K. L. Wang, Appl. Phys. Lett. 49, p. 847, 1986.

15. Li, X., Abe, T., Liu, Y., and Esashi, M., J. <u>Microelectromechanical Systems,</u> 11, 625-630, (2002).

16. Todd, S.T., Huang, X.T., Bowers, J.E., and MacDonald, N.C., J. Microelectromechanical Systems, 19, 55-63 (2010).

17. Davarik, B., Koburger, C.W., Schulz, R., Warnock, J.D., Furukawa, T., Jost, M., Taur, Y., Schwittek, W.G., DeBrosse, J.K., Kerbaugh, M.L., Mauer, J.L., <u>Electron Devices Meeting, 1989</u>, 61–64 (1989).

18. Oliver, M.R. (Ed.), Chemical-Mechanical Planarization of Semiconductor Materials, Springer-Verlag, Berlin Heidelberg, 2004.

Materials challenges for water supply

Water overlayers on Cu(110) studied by van der Waals density functionals

Sheng Meng

Beijing National Laboratory for Condensed Matter Physics, and Institute of Physics, Chinese Academy of Sciences, Beijing, 100190, China

Abstract: We investigate water overlayer structures on a model open metal surface: Cu(110), employing recently developed van der Waals density functionals (vdW-DFs). Both intact and half-dissociated layers are considered. We found that all the three structures (H-up, H-down and half-dissociated layer) have very close adsorption energies (differences < 0.16 eV) in all density functionals used here (two GGAs and two vdW-DFs), implying they have similar wetting orders on this surface. More importantly, we found that a hybrid vdW-DF approach treating water-Cu interaction with PBE exchange and hydrogen bonds with revPBE exchange yields best results for explaining experimentally observed wetting behavior of water on Cu(110).

1. Introduction

Copper, as one the most commonly used metal, play importance roles in many industrial processes and technological applications. For example, it serves as a cheap and most efficient catalyst for water-gas-shift reaction (H_2O $+CO{\rightarrow}H_2+CO_2$), for which the exact microscopic mechanisms are yet not known [1]. Water adsorption and dissociation are believed to be the rate-limiting process [1]. Another example is the usage of Cu container for safe storage of nuclear waste containment for over 100,000 years. However, a recent research has challenged this view, whereby it is shown that at the presence of liquid water, the lifetime of Cu films are much shortened; as a result, Cu films at least 1 m thick is required [2]. All these findings point to that understanding water adsorption and interaction with Cu surfaces is very important and much desired, especially concerning the microstructures of the water layers at the interface region.

The interaction of water with metal surfaces has received intensive investigations in past three decades [3], most on the closely-packed noble metal and transition metal surfaces and a few on open metal surfaces such as Cu(110). Recently, experimental [4,5] and theoretical efforts [6,7] have identified Cu(110) as another important borderline case that separates intact and dissociative water adsorption, along with Ru(0001) [8], based on the small energy differences between intact and dissociative water overlayers. Recently, a new (7×8) phase of water overlayer structure is observed in low energy electron diffraction (LEED) experiment, besides the commonly observed $c(2\times2)$ patterns [9]. The new phases was assigned to intact water adsorption with a mixture of H-down (67±15%) and H-down bilayers (33%); while the $c(2\times2)$ patterns were originally thought coming from either intact adsorption or water dissociation [4,5,7]. Most recently, other interesting structures, e.g. parallel water chains separated by ~50 Å were also observed [10, 11], and assigned to the formation of a chain of water pentagons based on density function theory (DFT) calculations and scanning tunneling microscopy (STM) imaging [11].

The appearance and dominance of water overlayer structures on Cu(110) is determined by a delicate balance between water-surface interactions and hydrogen bonding among water molecules. The competition between the two interactions determines the relative stability, characterized by water adsorption energy, of difference water structures. Previous studies have generally used DFT with generalized gradient approximation (GGA), presumably, PBE functional form [12], to study water adsorption on Cu(110). The obtained energies for water overlayers are generally smaller than the cohesive energy of bulk ice Ih [8], thus unable to explain the observed thermodynamic wetting layer. Moreover, the van der Waals (vdW) forces, which could play a larger role on open metal surfaces, are not included in PBE [13]. Here we examine the energy stability of water layers on Cu(110) employing recently developed vdW density functional (vdW-DF) approach [13], which treats the nonlocal vdW forces nicely for disperse systems [14]. By comparison to data from experiment and higher-level theory, we found that PBE overestimates hydrogen bonding energies in ice, while underestimates water-Cu interactions. We argue that a hybrid vdW-DF approach, with revPBE exchange for hydrogen bonding and PBE exhange for water-metal binding, offers a better description of water overlayer energies on Cu(110).

2. Method

The vdW interaction is included in the nonlocal correlation functional proposed by Langreth, Lundqvist and collaborators [13]. The exchange-correlation energy of vdW-DF is expressed as,

$$E_{xc}^{vdW} = E_x^{GGA} + E_c^{LDA} + E_c^{nl} ,$$

where E_x^{GGA} is the GGA exchange energy, and E_c^{LDA} and E_c^{nl} is the local (using the common local density approximation, LDA) and nonlocal part of the correlation energy. The latter include contribution from vdW energies, and is constructed as integration of electron densities at r, and r', with an interaction kernel $\phi(r,r')$,

$$E_c^{nl} = \frac{1}{2}\iint dr dr' n(r)\phi(r,r')n(r') .$$

To match the parameters in E_c^{nl}, the original choice of E_x^{GGA} for describing disperse interactions is given as revPBE [15] by Langreth et al., however, many other forms of the exchange functional have been tested and optimized [16,17] to yield a better match to the data from higher-level theory (MP2 or CCSD(T)) for a set of important dimers (S22 dataset) [18]. In this work, we are particularly interested in the original revPBE exchange functional (termed as vdW-DF/revPBE) and PBE exchange form (vdW-DF/PBE); the latter has been shown to give a closer match to MP2 results than revPBE for molecular dimer interactions.

The above scheme has been implemented in Siesta program, based on a self-consistent electron density calculation [19], which is different from the plane-wave non-self-consistent implementation [20]. We use pseudopotentials of the Troullier-Martins type [21] to model the atomic cores, and local basis set of double-ζ polarized orbitals for O (13 orbitals) and H (5 orbitals). For Cu, we use the basis set optimized for surface calculations by Garcia-Gil et al. [22], namely, double-ζ for 4s and 3d orbitals with an additional diffusive orbital (radius: 7.0 a.u.) for surface layers. An auxiliary real space grid equivalent to a plane-wave cutoff of 100 Ry is used for the calculation of the electrostatic (Hartree) term throughout. Calculated lattice constant of 3.63 Å is close to experimental value 3.61 Å. The small $c(2\times2)$ unit cell and a slab with six atomic Cu layers are adopted to model the intact water layers in the larger (7×8) arrangement and the dissociated layer in $c(2\times2)$. Previous studies show absorption energy differs <30-50 meV if a larger unit cell is used [8,23]. A vacuum layer of thickness exceeding 13 Å is included. A k-point mesh of $(2\times2\times1)$ guarantees adsorption energy convergence of 0.005 eV as compared to k-$(6\times6\times1)$ results. For structure optimization, we use the PBE form of the exchange-correlation functional and an adsorption structure is considered fully relaxed when the magnitude of forces on the atoms is smaller than 0.04 eV/Å Atomic structures optimized with revPBE are very close to those with PBE, for instance, the water heights in H-down layer are 2.350 Å and 3.044 Å in revPBE, as opposed to 2.343 Å and 3.055 Å in PBE. For consistence and easier detection of the influence of various functionals, we chose the PBE structures. For all energies reported the basis-set superposition error is excluded by the counter-poise scheme.

3. Results and discussion

Figure 1 displays three standard structure models considered for water overlayers on Cu(110): H-up layer, H-down layer and half-dissociated layer. For the first two structures water forms a puckered hexagonal network with every water having three hydrogen bonds to its immediate neighbors; the uncoordinated OH group for every two water molecules either pointing down to (H-down) or up away from surface (H-up) [6,7]. If this OH bond is broken and the dissociated H atom adsorbs directly on surface metal atoms, the structure is referred to as half-dissociated overlayer. The structures we calculated here are very close as reported before [6,7]: the structural parameters including the O-Cu bond lengths, hydrogen bond lengths, and O-O vertical thickness, are listed in Table 1. After a half of water molecules dissociate, the water "bilayer" with a OO vertical distance of 0.36 Å (H-up) and 0.31 Å (H-down) becomes planar, the resulting thickness is 0.12 Å; at the same time the bond length of O-Cu shrinks after dissociation. Similar trends

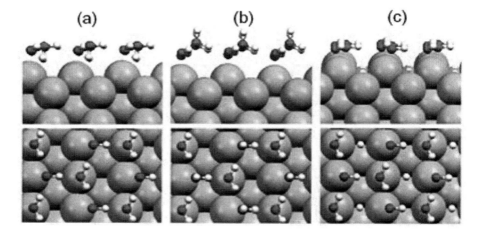

are observed for water layers on Ru(0001) [8].

Table 1. Structural parameters for water overlayers on Cu(110). Bond lengths (d_{O-Cu}, d_{O-O}) and vertical distances (z_{O-O}) are in unit of Å.

	H-down	H-up	Half-dissociated
d_{O-Cu}	2.34, 3.06	2.50, 3.24	2.05, 1.95
d_{O-O}	2.64, 2.78	2.59, 2.81	2.85, 2.69
z_{O-O}	0.31	0.36	0.12

We are more interested in the energetics of these overlayers with and without van der Waals interactions taken into account. Table II lists the adsorption energies calculated from two GGAs (PBE and revPBE) and two corresponding vdW-DFs. The adsorption energy (E_a) is defined as

$$E_a = \{E[slab] + n \times E[H_2O] - E[(H_2O)_n / slab]\} / n,$$

where $E[slab]$, $E[H_2O]$, and $E[(H_2O)_n / slab]$ are total energies of the bare Cu(110) slab, a single free water molecule, and the adsorption system, respectively. n is the number of water molecules in the super cell ($n = 4$ for $c(2 \times 2)$).

It is clearly seen from Table 2 that for all exchange-correlation functionals used here the H-down layer has the largest absorption energy among the three overlayer structures. The energy difference between H-down and H-up layers is 0.08, 0.06, 0.06, 0.03 eV, in PBE, revPBE, vdW-DF/PBE and vdW-DF/revPBE, respectively. The inclusion of ver der Waals interactions decreases the energy differences between the two intact structures, largely because in the H-up layer the van der Waals attraction between the upper water and the Cu surface not counted by any means in GGA (even not via OH...Cu bonding as in the H-down layer) is now included. The energy differences between H-down and half-dissociated layers are larger, being 0.11, 0.14, 0.10, 0.16 eV for PBE, revPBE, vdW-DF/PBE, and vdW-DF/revPBE, respectively. In the current approach using local bases, intact layers have a higher adsorption energy, at variance with the plane-wave results with sparse k-point sampling [6]. Nevertheless, for all functionals used here the energy difference between intact and dissociative adsorption does not exceed 0.16 eV per H_2O, indicating that the two structures are close in energy and Cu(110) is a borderline case for intact and dissociative water adsorption [6]. We also notice that the revPBE functional generally produces the binding energies too low to be reasonable for water adsorption, almost at the half value of that from PBE functional. Consequently, vdW-DF employing a revPBE exchange also yields lower energies, as compared to PBE and vdW-DF/PBE.

Table 2. Adsorption energy (in eV) of water overlayers on Cu(110) in different density functionals.

Functional	H-down	H-up	Half-dissociated
PBE	0.547	0.468	0.437
revPBE	0.280	0.218	0.145
vdW-DF/PBE	0.675	0.614	0.579
vdW-DF/revPBE	0.431	0.397	0.272
Hybrid*	0.565	0.503	0.480

*See text for definition.

More importantly, all the reported energies for water overlayer structures are lower than the corresponding cohesive energy (E_{ice}) of bulk Ice Ih. Employing a 12-molecule Hamann's model [24] for Ice Ih, we have calculated the cohesive energy of ice is 0.638 eV in PBE, 0.741 eV in vdW-DF/PBE and 0.560 eV in vdW-DF/revPBE. With these numbers it means thermodynamically no water overlayers considered above would wet the Cu(110) surface, at variance with experimental observation [9]. The most stable structure (the H-down layer) has an adsorption energy 0.09 eV in PBE, 0.07 eV in vdW-DF/PBE, and 0.13 eV in vdW-DF/revPBE lower than the E_{ice}, respectively, therefore cannot account for the wetting phenomenon in experiment. Note that water in (7×8) arrangements would have a slightly higher absorption energy (by 0.03-0.05 eV) than in $c(2\times2)$ if a similar hexagonal water network is assumed [8,23]; however, the large size of this structure prohibit an exact examination from first-principles.

We then try to analyze the errors associated with these computation approaches, and even better, if we can resolve the apparent discrepancy between the theoretical energetics of these water overlayers and experimental observations. First, we compared the ice cohesive energy E_{ice} obtained from these functionals to the experimental value [25], 0.58 eV, and reference data from high-level quantum chemistry calculations, 0.57 eV in CCSD(T) and 0.58 eV in MP2 [26], we conclude that PBE functional overestimates hydrogen bonding energies in bulk ice; the closest match to experimental and theoretical reference data is the one employing vdW-DF/revPBE (0.56 eV). Second, we also analyze possible errors in water-Cu interactions. It is known that PBE underestimates weak interactions such as dispersive adsorption of aromatic molecules on surfaces [14], we argue here that it possibly also underestimates the water-metal binding energy, which is quite weak too (0.1-0.3 eV). Despite the fact that there are rarely precise experimental or high-level theoretical data available on water-metal binding, we found that the above argument is supported at least in the case of single-water adsorption on NaCl, where the water-Na binding energy of 0.487 eV is calculated in CCSD(T) [27]. PBE gives only 0.328 eV as the binding energy [16]. We further extend to find that even vdW-DF with revPBE exchange underestimates water-metal interactions: this is the case since vdW-DF/revPBE even produces smaller adsorption energies than PBE for many configurations of single water on Cu(110). This is also supported by calculations on water/NaCl: vdW-DF/revPBE produces a binding energy [16] of only 0.334 eV, close to PBE result and significantly smaller than the CCSD(T) value. Therefore as an alternative we choose vdW-DF/PBE to better account for water-Cu interactions, which are much underestimated in other functional forms.

We propose a hybrid vdW-DF approach for obtaining adsorption energies of water structures on metal surfaces, namely, to use vdW-DF/revPBE for describing hydrogen bonds in ice adlayers (since it produces the best result for ice Ih) and vdW-DF/PBE for water-Cu interactions to remedy the underbinding problem in other functionals. In practice, we calculate the adsorption energy of a single water molecule in the same configuration as in the water overlayers using both vdW-DF/PBE and vdW-DF/revPBE, their energy difference ΔE_a is then added into the vdW-DF/revPBE adsorption energy of the corresponding overlayer on surface. In doing this we assume that the water-Cu interaction is not much disturbed by the presence of hydrogen bonds between water molecules; and can be separated from hydrogen bonding interactions. Effectively, this amounts to the vdW-DF/PBE treatment of water-Cu bonds, and vdW-DF/revPBE of hydrogen bonds for water overlayer adsorption on surfaces, as formulated above.

Table 3 lists the adsorption energies of a single-water molecule in the water overlayers in PBE, vdW-DF/PBE and vdW-DF/revPBE functionals. It is shown that vdW-DF/revPBE results are quite similar to PBE results, but both are much smaller than vdW-DF/PBE values. The differences between the two vdW-DFs are added to the vdW-DF/revPBE adsorption energy of water overlayers on Cu(110), as also listed in the last row of Table 2. The numbers (H-down: 0.565 eV; H-up: 0.503 eV; half-dissociated: 0.480 eV) are now much closer to the cohesive energy of ice E_{ice} (0.56 eV within vdW-DF/revPBE). These energy numbers are plotted as bars in Fig. 2. In particular, the H-down layer has a larger adsorption energy than E_{ice}, indicating that it wets the Cu(110) surface thermodynamically, as observed in experiment [9]. The (7×8) structure with the majority of H-down water configuration is observed to be a wetting layer on Cu(110). In reality a mixture of H-up and

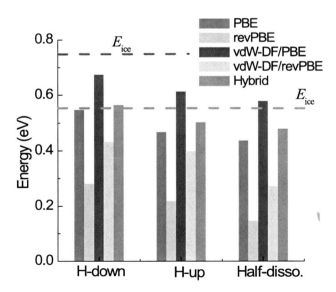

Figure 2. Adsorption energies of water overlayer on Cu (110) in various density functionals the hybrid approach. Dashed lines indicate the cohesive energy of ice Ih in vdW-DF/revPBE (magenta)

H-down layers in a larger periodicity will have a larger adsorption energy, confirming further its wetting ability. In addition, our results seem to suggest that the wetting behavior of half-dissociated layer may have a different origin: the hydrogen atoms from dissociated water molecules will easily diffuse away along [110] grooves forming hydrogen molecules and being desorbed; the mixed $OH+H_2O$ layer that is left has a very large binding energy to $Cu(110)$ (considering free OH and H_2O as references). Consequently, the half-dissociated layer is highly stable and wetting.

Table 3. Single-water adsorption energy (in eV) for water overlayers on $Cu(110)$ calculated from different density functionals. Labels "water1" and "water2" represent the two nonequivalent water molecules in each unit cell for the overlayer structures. ΔE_a is the energy difference between vdW-DF/PBE and vdW-DF/revPBE results, which is added to the total adsorption energy.

	H-down		H-up		Half-dissociated	
	water1	water2	water1	water2	water1	water2
PBE	0.272	0.109	0.280	0.101	0.031	-0.437
vdW-DF/PBE	0.370	0.236	0.375	0.250	0.146	-0.366
vdW-DF/revPBE	0.220	0.118	0.233	0.182	-0.087	-0.549
ΔE_a	0.150	0.118	0.142	0.069	0.233	0.183

4. Conclusions

We have investigated the energetics of model water overlayers on $Cu(110)$ employing density functionals with or without van der Waals interactions taken into account. Within all these functionals the intact and half-dissociated water layers have close adsorption energies with a small difference <0.16 eV. We found that PBE overestimates hydrogen bonding strength in ice and underestimates water-metal bonds; as a remedy a hybrid vdW-DF approach by treating hydrogen bonds with revPBE exchange and water-Cu bonds with PBE exchange has been proposed, which yields best energetics resolving the discrepancy with experimental observation of water wetting. This work sheds light on further development of more accurate density functionals for describing water interactions with solid surfaces.

Acknowledgments: The author acknowledges active discussions with Dr. I. Hamada and financial support from CAS.

Water overlayers on Cu(110) studied by van der Waals density functionals 75

[1] C. Callaghan, I. Fishtik, R. Datta, M. Car and A. Lugo, Surf. Sci. 541, 21 (2003).
[2] G. Hultquist et al., Cata. Lett. 132, 311 (2009).
[3] See, for example, in a recent review : A. Hodgson and S. Haq, Surf. Sci. Rep. 64, 381 (2009).
[4] Ch. Ammon, A. Bayer, H.-P. Steinr̈ck, and G. Held, Chem. Phys. Lett. 163, 377 (2003).
[5] K. Andersson, A. Gomez, C. Glover, D. Nordlund, H. Ostrom, T. Schiros, O. Takahashi, H. Ogasawara, L. G. M. Pettersson, and A. Nilsson, Surf. Sci. 585, L183 (2005).
[6] J. Ren and S. Meng, J. Am. Chem. Soc. 128, 9282 (2006).
[7] J. Ren and S. Meng, Phys. Rev. B 77, 054110 (2008).
[8] P. J. Feibelman, Science 295, 99 (2002).
[9] T. Schiros, S. Haq, H. Ogasawara, O. Takahashi, H. Ostrom, K. Andersson, L. G. M. Pettersson, A. Hodgson, and A. Nilsson, Chem. Phys. Lett. 429, 415 (2006).
[10] Yamada, S. Tamamori, H. Okuyama, and T. Aruga, Phys. Rev. Lett. 96, 036105 (2006).
[11] J. Carrasco, A. Michaelides, M. Forster, S. Haq, R. Raval, A. Hodgson, Nature Mater. 8 (2009) 427.
[12] J. P. Perdew, K. Burke, and M. Ernzerhof, Phys. Rev. Lett. 77, 3865 (1996).
[13] M. Dion et. al., Phys. Rev. Lett. 92 246401 (2004).
[14] D. C. Langreth et. al., J. Phys.: Condens. Matter 21, 084203 (2009).
[15] Y. Zhang and W. Yang, Phys. Rev. Lett. 80, 890 (1998).
[16] J. Klimes, D. Bowler, and A. Michaelides, J. Phys.: Condens. Matter 22, 022201 (2010).
[17] A. Gulans, M. J. Puska, and R. M. Nieminen, Phys. Rev. B 79, 201105 (2009).
[18] P. Jurecka, J. Sponer, J. Cerny, and P. Hobza, Phys. Chem. Chem. Phys. 8, 1985 (2006).
[19] G. Roman-Perez and J. M. Soler, Phys. Rev. Lett. 103, 096102 (2009).
[20] I. Hamada, K. Lee, and Y. Morikawa, Phys. Rev. B 81, 115452 (2010).
[21] N. Troullier and J. L. Martins, Phys. Rev. b 43, 1993 (1991).
[22] S. Garcia-Gil, A. Garcia, N. Lorente, and P. Ordejon, Phys. Rev. B 79, 075441 (2009).
[23] S. Meng, E. G. Wang, and S. W. Gao, Phys. Rev. B 69, 195404 (2004).
[24] D. R. Hamann, Phys. Rev. B 55, R10157 (1997).
[25] E. Whalley, Trans. Faraday Soc. 53, 1578 (1957).
[26] A. Hermann and P. Schwerdtfeger, Phys. Rev. Lett. 101, 183005 (2008).
[27] B. Li, A. Michaelides, and M. Scheffler, Surf. Sci. 602, L135 (2008).

Challenges in conclusive, realistic and system oriented materials testing

Employment of high Resolution RBS to characterize ultrathin transparent electrode in high efficiency GaN based Light Emitting Diode

Grace Huiqi Wang[1] and Taw Kuei Chan[2].

[1] *Institute of Materials Research and Engineering (IMRE), Singapore.*
[2] *National University of Singapore (NUS), Physics Department, Singapore.*

Abstract

GaN based light emitting diodes (LEDs) have attracted significant attention for use in solid state lighting. The high efficiency of LEDs has provided substantial energy savings and environmental benefits in a number of applications. A thermally stable and highly transparent low resistivity electrode is important for the fabrication of high brightness LED. Technological advances in microelectronics, optoelectronics and photonics rely on the use of precisely manufactured thin film structures (<10nm) in the LED electrodes. Recent advent of nanotechnology demands the ability to fully and nondestructively characterize such films in the monolayer range. In this work, we present a new metallization scheme to reduce the contact resistance and enhance current injection into the LED for improved light extraction, using excimer laser irradiation. Various techniques, including Rutherford backscattering, transmission electron microscopy and UV-Vis spectroscopy techniques are used to study the contact properties of the thin film electrodes.

Introduction

GaN-based materials have received considerable attention for widespread applications in photonic and electronic devices[1,2]. The high efficiency of LEDs has provided substantial energy savings and environmental benefits in a number of applications.[3] Forming a low resistive, thermally stable and uniform ohmic p contact to wide bandgap semiconductors such as GaN is still a challenge. Contact properties limit the overall performance of optical and electronic devices. The performance of such LEDs and lasers remains limited by several problems related to the high resistance ohmic contact to the p-type GaN. Most of the reported methods for reducing p-contact resistance rely on the optimization of the contact annealing temperature and improvement of metal-semiconductor interface [4-6] and have shown improvement in the LEDs performance [7-9].

The conventional metal contact formed by rapid thermal annealing (RTA) induces thermal damage such as GaN decomposition and interfacial discontinuities[10], leading to spiky interfaces within LED structures which often results in higher contact resistivity[11]. In addition, low doping levels of p-GaN[12-15] further create difficulties in achieving a good ohmic characteristic, important to device performance. High temperature thermal annealing during device processing is widely used to remove the native oxide on the GaN surface[16-17]. The presence of the insulating native oxide at the metal/semiconductor interface results not only in poor interfacial structures but also in an increase of effective schottky barrier height to act as a barrier for the carrier to transport from the metal to the semiconductor [18-20].

In this paper, we propose to employ excimer laser irradiation to thermally anneal the Cr/Au contact formed on the p-GaN surface. Consequently, the resulting laser annealed contact on LED had improved light transmittance, thermal stability and surface morphology. Rutherford backscattering spectroscopy (RBS), with alpha particles as analytical probes, is the most versatile technique used for the nondestructive and quantitative depth profiling of the thin electrode. Conventional RBS is limited by energy resolution of the Si detectors used[21-22], the depth resolution is around 10nm for Au. Sub nanometer depth resolution is becoming a critical necessity and a challenge faced in material testing of the electrodes, thus, we employ a double focusing 90° sector magnetic spectrometer at the 45° beamline, to improve the depth resolution by an order of magnitude, allowing monolayer depth resolution to characterize the laser annealed electrodes for LED applications. The high resolution RBS facility with sub 1KeV energy resolution characterizes the ultrathin electrode and attempts to mitigate the challenges faced in today's materials testing.

The rapid miniaturization of electronic devices in recent years has caused the thicknesses of thin films in modern semiconductor research to be reduced to the regime of tens of angstroms. Conventional Rutherford Backscattering Spectrometry (RBS) can no longer provide depth profiling for thin films with such thicknesses. The High-resolution RBS (HRBS) offers thin film depth profiling and lattice strain analysis with sub-nanometer resolution through a more complicated detection system[21,23]. Energy detection involves a spectrometer magnet and a Micro-Channel Plate-Focal Plane Detector (MCP-FPD) assembly. The MCP-FPD utilizes an electron-multiplication process within micro-channels to amplify the signal generated by the incidence of every backscattering ion[21]. This work has attempted to investigate the effects of laser irradiation on the transparent contact electrode layers to lower the specific contact resistance and reduce the oxide formation on GaN.

Experimental details

The LED structure is grown epitaxially on the c-plane sapphire substrate comprising GaN/InGaN by metal organic chemical vapor deposition system

(MOCVD). Trimethylgallium, trimethylindium and ammonia are used as the precursors for Ga, In and N respectively. Nitrogen and hydrogen gas serve as the carrier gases for the growth of InGaN and GaN respectively. For the LEDs, multiple quantum well (MQW) structures with a number of periods are grown with InGaN well and GaN barrier on top of the Si doped GaN on sapphire template.After the growth of InGaN/GaN MQWs, an AlGaN layer is then deposited, which serves as an electron blocking layer prior to the top p-GaN growth. The mesa structures were patterned using reactive ion etcing (RIE) using BCl_3/Cl_2 at 6°C. The mesas were approximately 500nm deep and were used to prevent current spreading as well as to isolate contact pads. Fig. 1(a) details a cross section illustration of the GaN layers after mesa isolation. After patterning and definition of the p-GaN surface, fabrication of LEDs was carried out with the metallic contacts.

Prior to deposition of the metal films, all the samples were exposed to Ar plasma for 60s in the RIE chamber. Then Ti (15nm)/Al(220nm)/Ni(40nm)/Au(50nm) structures were deposited by electron beam evaporation. The samples were then annealed at 900°C for 30s in N_2 ambient. Transparent p contact electrodes Cr(3nm)/Au (3nm) or Ni(3nm)/Au(3nm) were then deposited by electron beam evaporation. Au(200nm) p bond was then defined by lithography to contact the p contacts and reduce the contact resistance and enable p electrode probing. As a first demonstration in examining the impact of laser annealing in enabling an enhancement in hole injection current, various laser annealing conditions were explored. Single and multiple laser pulses at various laser excitation energies in the range of 20 to 500 mJ cm^{-2} are considered. Below 380mJ cm^{-2}, the laser energy could not sufficiently achieve p-ohmic characteristic in the p contact. Hence, RTP at 575°C 30sec was further employed to obtain an ohmic characteristic in the electrode.

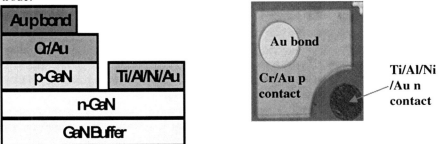

Figure 1: Schematic cross sectional views of light emitting diode layers after mesa definition. Optical micrograph of fully processed LED after p bond formation is shown.

Following the fabrication, the processed LED devices were characterized and the current-voltage (*I-V*) plots from the laser annealed and RTP contacts were obtained. Fig. 1(b) shows an optical micrograph of a typical LED with an emission area of 300 by 300μm^2. Fig. 2 shows the X-ray diffraction (XRD) rocking curve of the GaN/InGaN LED layers on bulk Si. The GaN (0002), AlN(0002), buffer

AlGaN (0002) reflections and well defined superlattice fringes created by the In-GaN/GaN multi-quantum wells are shown.

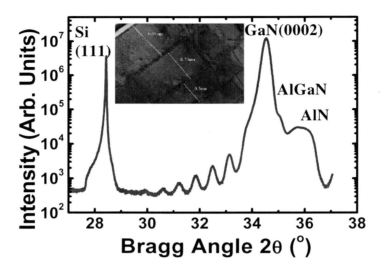

Figure 2: HRXRD spectra of the LED structures grown on bulk Si(111) substrates. The superlattice fringes are clearly resolved and represent high crystalline quality of gallium nitride template for LED fabrication

Rutherford Backscattering Spectrometry

During a typical RBS measurement (Fig. 3), incident ions of energy $E0$ are backscattered into a detector placed at a certain scattering angle θ, which measures the backscattered ion energy $E1$. The spectra obtained, after a series of electronic signal processing, is a number distribution of backscattered ion energies $E1$, which is then used to estimate the composition of the target, by numerical simulations of the energy spectra with the software package SIMNRA[24].

For thin contact electrodes deposited on thick substrates, we are mainly interested in the stoichiometry and depth profile and interdiffusion mechanisms of heavy elements within the films at the interface region with the substrate.

Overview of HRBS endstation

The HRBS endstation was fabricated by the Machinery Company of Kobe Steel Ltd and installed as shown in Fig. 4. The general setup consists of:
  i.   Main chamber with a load lock chamber
  ii. UHV Vacuum sytem (pumps, valves and interlocks)
  iii. 5-axis goniometer
  iv. Control cabinet
  v.   Spectrometer magnet and MCP-FPD chamber (HRBS detection system)

Both the main and the MCP-FPD chambers are constantly maintained under UHV with two turbo-molecular pumps, which are located beneath the main chamber and the MCP-FPD chamber. Sample exchange is performed by a transfer rod which transfers a target holder onto the goniometer attachment in the main chamber from a load-lock chamber through a gate valve. A controller program at the control cabinet oversees the vacuum interlocks system and allows for programmed or manual control of all valves as well as the goniometer rotation axes.

During measurements, the divergence of the ion beam is defined using the motorized slits located ~ 1 m before the main chamber. The scattering target is suspended by a precision 5-axis goniometer and is applied with an electrical potential for secondary electron suppression. Backscattered ions subsequently enter the detection system, which consists of a 90° spectrometer magnet and a Micro-Channel Plate – Focal Plane Detector (MCP-FPD) stack. The system is mounted on a circular track which allows the scattering angle to be varied via 7 detection ports. The signal output from the FPD is processed by a network of electronics on the control cabinet to extract the position of incidence of the ion along its length. The spectrum is sorted by an MCA before being output by software on a computer.

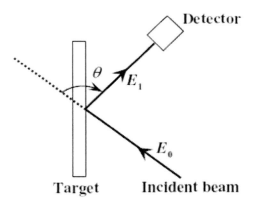

Figure 3: General layout of typical RBS measurement

Figure: 4 Layout of the 45° beamline and the HRBS endstation.

HRBS Detection system

Superior energy resolution of HRBS is achieved by replacing the solid-state PIPs detector in normal RBS with a detection system that consists of a spectrometer magnet and a 100 mm long by 15 mm width Micro-Channel Plate (MCP) – Focal Plane Detector (FPD) stack. Backscattered ions of energy E enter the spectrometer magnet at a specific scattering angle where they follow circular trajectories with radii r that are proportional to $E^{0.5}$(Fig 5). These ions are then incident onto the MCP-FPD stack at the focal plane of the spectrometer.

The incidence of a backscattered ion onto the MCP triggers an electron cascade, which deposits a charge/voltage pulse onto the FPD. The FPD, being of a resistive strip type, determines the position of the pulse and hence the ion incidence along its length using a simple charge division method. The spectrum obtained directly from the system is therefore a "position spectrum", which is a position distribution of backscattered ions at fixed intervals of length along the MCP-FPD.

Backscattering study of thin electrodes

Rutherford backscattering (RBS) was employed to study the diffusion behaviors of the transparent contact metals in an area of 1-2mm^2. The distinct advantage of Rutherford Backscattering Spectrometry (RBS) is its ability to provide quantitative and non-destructive depth profiling of the elements within the sample[21].

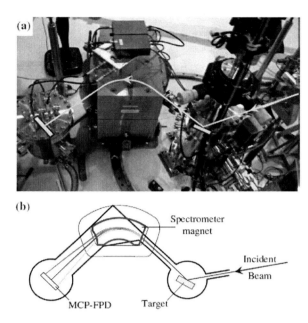

Figure 5: (a) The actual and (b) schematic layout of the HRBS detection system. Ions backscattered from the target at a fixed scattering angle enters the spectrometer magnet. Ions with higher momentum will be bent with a larger radii (red) while lower momentum ions will have a smaller radii (green).

As researchers constantly attempt to improve the luminescence properties of LEDs, this process has reached the point where the required thickness of p contact has reached nanometer regime. The superior depth resolution of High resolution RBS coupled with its non-destructive and quantitative spectrometric capability makes it highly suitable to probe the elemental profiles of ultra thin electrodes. Of particular interest is the interface region between the metal contact and the GaN substrate, since the contact-semiconductor interface quality will affect the electrical and morphological performance of the LEDs. The strain imposed by the thin electrode on the GaN lattice at the interface will also affect the carrier mobility in the p-n junction of the LED and affect LED brightness.

The Cr/Au on GaN samples were measured using a 500 keV He^+ beam which was collimated to 2 by 2 mm size with a divergence angle of less than 1 mrad. The beam was incident on the Cr/Au sample and ions scattered at 65° were analyzed by the spectrometer and were collected by the MCP-FPD. Non-aligned (random) HRBS spectra of these samples were measured at a sample tilt of 40° under the IBM geometry, which were then simulated and fitted using the SIMNRA program to obtain the elemental depth profiles.

Fig. 6 shows the RBS spectra from Cr(5nm)/Au(5nm) metal and Fig. 7 shows that of Cr(3nm)/Au(3nm) metal on GaN and after RTP and laser annealing respectively. The channel number 410 corresponds to Ga at surface respectively. The lines labeled Au, Cr and Ga indicate the energies of the He$^+$ ion backscattered from the surface of the respective Au, Cr and Ga layers. The heavier the atoms, eg. Au, the higher the energies of the backscattered He$^+$ ions. Therefore, the broad peak on the extreme right is mainly contributed by Au signal. The N signals from GaN also pile up on the Ga signals on the lower energy side. However the backscattering signals of N atoms are wekeaer due to their smaller differential scattering cross section. Therefore, the signals of the different atoms can be easily distinguished from one another by the energies of the backscattered He$^+$ ions. In this case, the shape of an element's profile will reflect the diffusion behavior or the element. The analytical characterization of the contacts is studied using RBS utilities and manipulation package (RUMP) simulations [25-26].

Comparing RTP and laser annealing, RTP causes the Cr thickness to increase and the maximum yield backscattered from Cr atoms reduce. This shows that probably CrN or CrO layers were created, since the interface between Cr and GaN moved into the GaN layer. With increasing laser energy, the Cr signals extend to the lower energy side. Therefore the extension of the Cr signals is an indication of the interdiffusion of Cr. Using HRBS, it was found the strain is largest at regions next to the interface in all samples, and decreases in magnitude with increasing distance from the interface, reducing to below detection limit at distances beyond

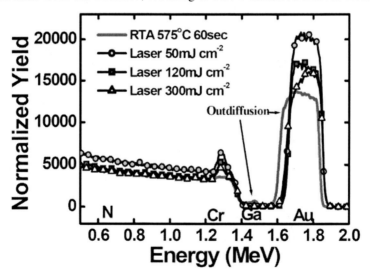

Figure 6: Rutherford Backscattering Spectra of Au(5nm)/ Cr(5nm)/GaN layer after undergoing rapid thermal annealing at 575°C, 60sec and laser annealing at various laser annealing conditions.

6 nm. The interface lattice strain is also found to be largest in the as-deposited sample, decreasing with samples annealed at increasing laser energies, so that the strain in the optimally laser annealed sample is below detection limit. Hence, contacts formed were in lattice compliance to GaN.

Employment of high Resolution RBS to characterize ultrathin transparent electrode 87

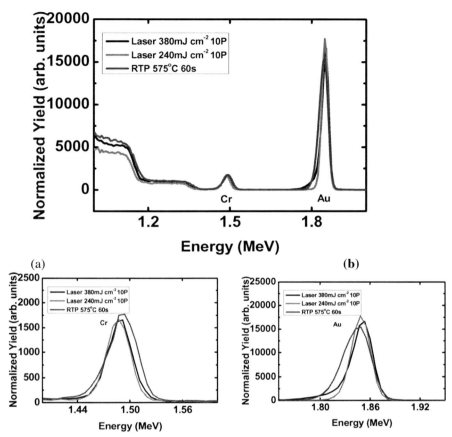

Figure 7 Rutherford Backscattering Spectra of Au(3nm)/ Cr(3nm)/GaN layer after undergoing rapid thermal annealing at 575°C, 60sec and laser annealing at 0.3Jcm−2. (a) and (b) is the zoomed in view of Cr and Au respectively.

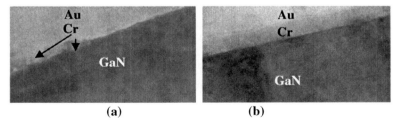

Figure 8: Bright field TEM images of Cr/Au/GaN layer after undergoing (a) rapid thermal annealing at 575°C, 60sec (b) laser annealing at 0.38Jcm−2, 10 pulses. Rapid thermal annealing shows the development of compositional segregation due to the consumption of Cr and diffusion of Cr, and laser anneal shows clearly defined interfaces remained, the Cr/Au interface can be discerned.

TEM image of Ti/Al Formation on gallium nitride

Bright field cross sectional TEM images prepared from Cr/Au contacted samples after undergoing (a) RTP and (b) Laser Annealing, are illustrated in Figure 8. TEM is performed to investigate the morphological changes in Cr/Au. After undergoing laser annealing, the Cr and Au interface remains clearly visible, but when RTP was performed, the Cr/Au interface is unclear. A significant change in the integrity of the contact/semiconductor interface was observed for samples thermally annealed at or above 575°C. For laser annealed samples, the contact shows an abrupt and pristine interface, and laser annealing causes minimal or no penetration of material further into the GaN/InGaN.

The interface between the contact electrode materials and semiconductor are not degraded when laser irradiation using multiple (1, 5, 10 & 20) pulses was performed. This shows that the interfaces between the contact materials and GaN are not affected by the short duration, high intensity laser annealing conditions.

Impact of laser annealing on SIMS profiles

SIMS data in figure 9 (a) and (b) further reveal an increased consumption of Cr and penetration of Au into Cr, with RTP. Also, a steeper elemental profile reveals a more abrupt junction formed. Thus laser annealing reduced Ga, Au and Cr diffusivities, thus minimized surface roughening issues. However, RTP caused a gradual diffusion of Ga, Cr and caused significant intermixing between the atoms. The oxygen content was also reduced with laser irradiation. The reduction of the insulating native oxide at the metal/semiconductor interface (~6nm from top surface) leads to improved interfacial properties and reduced the effective Schottky barrier height for easier carrier transport across the metal–semiconductor interface.

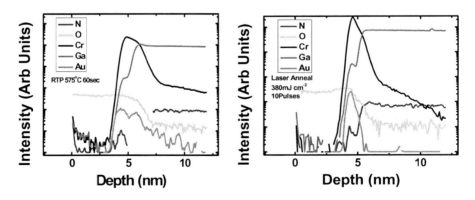

Figure 9. Comparison of elemental depth profiles after undergoing RTP at 575°C, 60sec and laser annealing at 120mJ cm-2. Au and Cr indiffusion and Cr and Ga out-diffusion were observed after RTP shown by the long diffusion tail in the SIMs profiles.

Contacts transparency and light transmission properties

Transmitting properties of p contacts on GaN were further studied in ultraviolet and visible region by using a UV–VIS spectrophotometer equipped with an integrating sphere[27]. The test for each specimen was conducted for three times at different sites selected randomly. The three test results were almost identical, showing a good homogeneity of the laser annealed and RTP p contacts on GaN and one of the results was chosen for discussion in this article. The curves of the transmission spectra consist of smooth undulations and sharp peaks. The undulations are caused by interference between the lights reflected from the film surface and the film/substrate interface. The peaks of the waves are decided by the optical properties of the films and substrates, as well as the thickness of the films. The sharp peaks were caused by the absorption of the films. For transparent or weak absorbing thin films, the envelopes of the maxima and minima in the transmission spectrum are continuous and smooth.

There was no significant spectral shifts whether laser annealing or RTP was employed in the contact annealing as shown in Fig.10. The transmission properties of laser annealed Cr-Au at 450 mJ cm^{-2} was slightly higher than RTP in the 470nm blue light region. Cr-Au achieved close to ~80% transmissivity at λ of 470nm after undergoing laser irradiation at 450 mJ cm^{-2}. Comparing the various laser annealing condition, it was observed that, increasing laser irradiation achieved improved optical transparency in the laser annealed contacts. Laser annealing at an optimal energy could even surpass the transmission properties of a RTP contact. When the RTP temperature was increased from 400°C to 575°C, a similar trend was observed. The overall transmission properties of the contacts were enhanced at an elevated annealing temperature.

Similarly, the transmission properties of Ni-Au peaked at a laser annealing condition of 240 mJ cm^{-2}. Ni-Au achieved close to ~62% optical transmission at an optimal laser anneal condition. Similar levels of optical transmission were observed when the laser energy increased from 200 to 240 mJ cm^{-2}. In comparison, Cr- Au has a much higher irradiation resistance than Ni-Au and Cr-Au achieved slightly improved light transmission at a wavelength of 470nm as compared to Ni-Au contacts.

Electroluminescence properties

After the LED was fully processed, the emission from the sample with laser annealed and RTP contacts were testing using room temperature electroluminescence (EL) mapping. For the LED with sapphire substrate, bright green emission was clearly observed from LEDs at a wavelength of 510nm, with an injection current of 50mA [Fig. 11]. Fig. 11 also shows the room temperature electroluminescence spectra taken from the green LED with increasing injection current. The injection current was gradually increased from 10mA to 70mA. The light output from both LEDs increase linearly with injection current.

The EL Intensity of the LED sample with laser annealed contacts is twice as high as that of a LED with RTP contact. The laser annealed regions gave stronger luminescence than the conventional p-GaN regions formed by RTP alone. Laser annealing helped to achieve improved ohmic characteristic, reduced contact resistivity and enhanced dopant activation of Mg dopants in the p region of the LED device. Based on the EL spectra shown in Fig.8, there is no prominent shift in wavelength when the injection current increased from 10mA to 70mA. Nevertheless, the slight shift was noted to be 512nm at 10mA to 507nm at 70mA. A similar phenomenon was also observed for the RTP sample. This is attributed to the incorporation of indium rich nano structures embedded in the InGaN well layer.

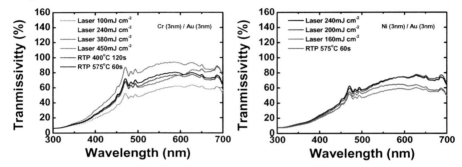

Figure 10: Visible light transmission spectra of specimens containing (a) Cr-Au on GaN and (b) Ni-Au on GaN, after undergoing various laser annealing and RTP conditions. Improved transmission is seen in contacts that have been laser annealed at a higher laser energy.

At low injection current, the electrons and holes recombine mainly at the InGaN quantum well (with higher indium composition). The graded indium composition of the InGaN well layer causes the electrons to and they recombine with holes to give bright green emission at 512nm for a current injection of 10mA. InGaN serves as effective carrier trapping sites in the InGaN as well as electrons having to overcome an additional potential barrier at the interface of the InGaN and GaN. Hence there is a greater tendency for carriers to thermalize in the InGaN. With a higher injection current, at 30mA or 50mA, the InGaN are now saturated with carriers and excess high kinetic energy carriers can overcome the potential barrier to diffuse to the GaN layer and give rise to green emission. Further increase in injection to 50mA or 70mA enables carriers to recombine radiatively at the GaN and InGaN interface, which results in a shift in the wavelength at higher injection currents to 507nm. Further increase in current, may even result in cyan coloured emissions. This may even enable a wavelength tunable LED to be achieved.

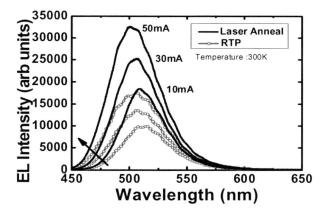
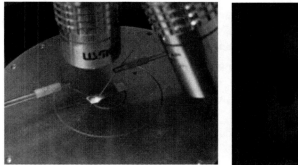
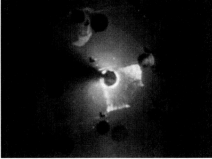

Figure 11. Twice as much improvement in EL intensity with laser annealing. Photographic images at injection current of 50 mA show bright green emissions. Red shift in the EL peak comparing RTP with laser annealed contact is attributed to slightly improved Mg activation in p-GaN. after undergoing various laser annealing and RTP conditions.

Laser annealing at elevated energies may roughen the LED surface and reduce the escape probability of photons due to the angular randomization of photons through internal scattering from the roughened sidewalls. As expected, this effect could enable an improvement in the light extraction efficiency of the LEDs.

Conclusion

In summary, ohmic contact characteristics for p type GaN in Ni/Au and Cr/Au contacts using excimer laser annealing have been obtained. The interaction between the metal and p-GaN which results in the formation of a GaN:Ni or GaN:Cr compound during the annealing was observed. Cr diffusion was promoted through laser annealing and the interaction resulted in ohmic characteristics in both the metallization schemes. The laser annealed contacts showed better ohmic characteristics due to uniform interaction profile after excimer laser annealing. Though Ni/Au showed improved ohmic characteristic, Cr/Au displayed greater tolerances towards enhanced laser energies employed, and could achieve ohmic characteristics similar to Ni/Au metallization. In summary, we have demonstrated the usefulness of employing laser annealing in thermally annealing the contacts. Laser annealing brought about several benefits in contact formation over RTP: (i) reduction of native oxide layer on p-GaN, and (ii) formation of a more intimate contact between metal and p-GaN.

In addition, an enhancement in electroluminescence intensity is observed from p-GaN with laser annealed transparent contacts. Electroluminescence spectra showed slight change of emission wavelength with increasing injection current. This is attributed to the carrier filling mechanisms in the graded InGaN quantum well (template of the LED). In conclusion, the HRBS is a highly useful spectrometric method to probe the composition and structure of thin films. The HRBS endstation works correctly and the accuracy of the output spectra is further enhanced by this work. Contact formation on GaN using pulsed laser irradiation could be timely to meet the demands for reliable ohmic contacts formation in the integration of III-V semiconductors for future device applications.

1. References

1. S. Nakamura, M. Senoh, N. Iwasa, and S. Nagahama, Jpn. J. Appl. Phys.,Part 2 34, L797 (1995).
2. T. Nishida, H. Saito, N. Kobayashi, Appl. Phys. Lett. 79, 711–712 (2001).
3. S.J. Pearton, C.B. Vartuli, J.C. Zolper, C. Yuan, R.A. Stall, Appl. Phys. D.J. King, L. Zhang, J.C. Ramer, S.D. Hersee, L.F. Lester, Mater. Res. Soc. Symp. Proc. 468 (1997) 421.
4. T. Mori, T. Kozawa, T. Ohwaki, Y. Taga, S. Nagai, S. Yamasaki,S. Asami, N. Shibata, M. Koike, Appl. Phys. Lett. 69 (1996) 3537.
5. T. Kim, J. Khim, S. Chae, T. Kim, Mater. Res. Soc. Symp. Proc. 468 (1997) 427.
6. C.H. Kuo, S.J. Chang, Y.K. Su, L.W. Wu, J.F. Chen, J.K. Sheu, J.M.Tsai, Jpn. J. Appl. Phys. 42 (2003) 2270.
7. C.H. Chen, S.J. Chang, Y.K. Su, Jpn. J. Appl. Phys. 42 (2003) 2281.
8. R.C. Tu, C.J. Tun, S.M. Pan, H.P. Liu, C.E. Tsai, J.K. Sheu, C.C. Chuo
9. T.C. Wang, G.C. Chi, I.G. Chen, IEEE Photon. Technol. Lett. 15 (2003) 1050.
10. Q.Z. Liu, S.S. Lau , Solid-State Electronics Vol. 42, No. 5, pp. 677-691, 1998
11. S. H. Liu, J. M. Hwang, Z. H. Hwang, W. H. Hung, H. L. Hwang, *Applied Surface Science*, *Volumes 212-213*, *Pages 907-911*, 2003.

12. F.B. Naranjo, E. Calleja, Z. Bougrioua, A. Trampert, X. Kong, K.H. Ploog, *Journal of Crystal Growth*, *Volume 270, Issues 3-4, Pages 542-546* , 2004.

13. G. C. Chi, C. H. Kuo, J. K. Sheu, C. J. Pan, *Materials Science and Engineering B*, *Volume 75, Issues 2-3, Pages 210-213*, 2000.

14. B. Schineller, A. Guttzeit, F. Vertommen, O. Schön, M. Heuken, K. Heime, R. Beccard , *Journal of Crystal Growth*, *Volumes 189-190, Pages 798-802*, 1998.

15. Jing-Bo Li, JingKui Liang, GuangHui Rao, Yi Zhang, GuangYao Liu, JingRan Chen, QuanLin Liu, WeiJing Zhang, *Journal of Alloys and Compounds*, *Volume 422, Issues 1-2, Pages 279-282*, 2006.

16. Heng-Kuang Lin, Hsiang-Lin Yu, Fan-Hsiu Huang, *Solid-State Electronics*, *Volume 54, Issue 5, Pages 552-556, 2010*.

17. Wu-Yih Uen, Zhen-Yu Li, Shan-Ming Lan, Tsun-Neng Yang, Yen-Wen Chen, Sen-Mao Liao, *Solid-State Electronics*, *Volume 51, Issue 3, Pages 460-465* ,2007.

18. T. Sawada, Y. Izumi, N. Kimura, K. Suzuki, K. Imai, S. -W. Kim, T. Suzuki, "*Applied Surface Science*, *Volume 216, Issues 1-4, Pages 192-197*, 2003.

19. Jae Wook Kim, Jhang Woo Lee, *Applied Surface Science*, *Volume 250, Issues 1-4, Pages 247-251, 2003*.

20. Yasuo Koide, H. Ishikawa, S. Kobayashi, S. Yamasaki, S. Nagai, J. Umezaki, M. Koike, Masanori Murakami, *Applied Surface Science*, *Volumes 117-118, Pages 373-379* ,1997.

21. T. Osipowicz, H.L. Seng, T.K. Chan, B. Ho, *Nuclear Instruments and Methods in Physics Research Section B: Beam Interactions with Materials and Atoms*, *Volume 249, Issues 1-2, Pages 915-917, 2006*.

22. P Aloupogiannis, A Travlos, X Aslanoglou, M Pilakouta, G Weber , *Vacuum*, *Volume 44, Issue 1,Pages 37-39* , 1993.

23. Kenji Kimura, Kaoru Nakajima, Michi-hiko Mannami, *Nuclear Instruments and Methods in Physics Research Section B: Beam Interactions with Materials and Atoms*, *Volumes 136-138, Pages 1196-1202, 1998*.

24. M. Mayer, SIMNRA User's Guide, Max-Plank-Institut Fur Plasmaphysik, Garching, Germany, 2006.

25. T.K. Chan, P. Darmawan, C.S. Ho, P. Malar, P.S. Lee, T. Osipowicz, Nuclear Instruments and Methods in Physics Research B 266 pp.1486, 2008.

26. P. Malar, T.K. Chan, C.S. Ho, T. Osipowicz, Nuclear Instruments and Methods in Physics Research B 266, pp. 1464, 2008.

27. Y. Yao, C. Jin, Z. Dong, Z. Sun, S.M. Huang, *Displays*, *Volume 28, Issue 3, Pages 129-132* , 2007.

A possible route to the quantification of piezoresponse force microscopy through correlation with electron backscatter diffraction

T. L. Burnett, P. M. Weaver, J. F. Blackburn, M. Stewart and M. G. Cain

National Physical Laboratory, Hampton Road, Teddington, TW11 0LW

Abstract

The functional properties of ferroelectric ceramic bulk or thin film materials are strongly influenced by their nano-structure, crystallographic orientation and structural geometry. In this paper, we present a possible route to quantification of piezoresponse force microscopy (PFM) by combining it with textural analysis, through electron back-scattered diffraction (EBSD). Quantitative measurements of the piezoelectric properties can be made at a scale of 25 nm, smaller than the domain size. The combined technique is used to resolve the effective single crystal piezoelectric response of individual crystallites in polycrystalline lead zirconate titanate (PZT). The piezoresponse results are quantified via two methods and these are compared to the piezoresponse predicted by a model. The results are encouraging for the quantification of the PFM technique and promote it as a tool for the future development of new nano-structured ferroelectric materials such as memory, nano-actuators and sensors. Knowledge of the orientation at the nanoscale allows for a method of quantification of the PFM signals and is being pursued as one method with which to potentially provide standards for PFM. This pre-standards work continues under a new VAMAS (Versailles Project on Advanced Materials and Standards) initiative. Standardization will provide a way to meaningfully compare values recorded with the PFM technique, which has become an important tool in the characterization of piezoelectric and ferroelectric materials.

Introduction

The history of standardization is a rich one. The first legal documentation regarding standardization came in England in 1196 AD with the 'Assize of Measures' which decreed "Throughout the realm there shall be the same yard of the same size and it should be of iron". Later standards dealt with the measurement of almost everything that was to be traded, for example in the Magna Carta 1215 AD

'There shall be standard measures of wine, ale, and corn (the London quarter), throughout the kingdom. There shall also be a standard width of dyed cloth, russett, and haberject, namely two ells within the selvedges. Weights are to be standardized similarly'.

Whilst individual countries attempted to control trade (and taxes) by standardizing measures it was not until the Meter Convention in Paris in 1875 that the first International Standard was established. The Système International d'Unitēs in 1960 built on this by providing the seven base units from which all other units are derived. [NB: The kilogram remains the last of these units to be still based on a physical artefact.]

Standards have come a considerable way since their beginnings but the impetus behind their establishment has not dwindled. It is only by establishing standards that quantities can be reliably compared, which is just as vital for research as for the measurement of commodities. Standards provide the ability to compare results effectively or to assure a certain thing will do a certain job (for example a resistor truly having the stated resistance (BS EN 60062:2005)).

Currently many different bodies govern different standards. The work in this project is an initiative set-up under the pre-standards committee, VAMAS (Versailles Project on Advanced Materials and Standards) TWA24 project 3. The work is intended to continue onto full standardization under ISO (International organization for standardization), although work has only just begun.

PFM was first described in 1992 [1] and since then has developed into an important tool for the characterization of ferroelectric and piezoelectric materials. Ferroelectric materials, which are also always piezoelectric, (although not all piezoelectric materials are ferroelectric) comprise an important class of materials. They have application in a wide range of devices from SONAR to ultrasound to diesel fuel injectors and a range of sensors such as accelerometers as found in mobile phones. The recent advent of FeRAM (ferroelectric random access memory) has driven a considerable amount of research with PFM being a key tool in this research area [2-3]. The ability to probe these types of material at the nanoscale is crucial for developing our understanding of the physics of these phenomena as well as characterizing materials and the devices made from them. Electrical domains (discrete regions with the same electrical polarization), which determine the electrical properties of these materials, exist at the nanoscale. Piezoresponse force microscopy (PFM) is currently the only technique that can directly measure the electromechanical coupling of piezoelectric materials with nanoscale resolution.

PFM measures the surface displacement of a piezoelectric material in response to an applied electric field. The field is applied through an electrically conducting atomic force microscope (AFM) probe. The surface displacement is the local response to the applied field via the converse piezoelectric effect and is measured by the probe. The resolution of the piezoresponse is approximately 1 pm, a resolution achieved through the use of an oscillating voltage and the use of a lock-in amplifier [4]. There is a focus of PFM work on thin films, and single crystals and the methodology has been extensively developed [5,6]. However a majority of PFM results are published as images without any quantification [2,3,7]. Alternatively when results have been quantified there is no consensus on the way that this should be achieved and a range of different methods have been employed [8,9].

The simplest and most commonly implemented PFM is so-called 'vertical' or 'out-of-plane' PFM, which measures the component of the piezoelectric response in the direction of the applied field i.e. perpendicular to the sample surface. Where $\Delta z(t) = \Delta z^{\pm}(t) \pm d_{zz} V_{AC} \sin(\omega t)$ and $\Delta z^{+}, P_z > 0$ or $\Delta z^{-}, P_z < 0$ [6]. Where $\Delta z(t)$ is the surface displacement, d_{zz} is the effective piezoelectric coefficients of the material and V_{AC} is the applied oscillating voltage. The polarization of the material positive or negative polarisation, P_z, dictates the direction of the surface displacement $\Delta z(t)$. That is to say there is a direct relationship between the piezoelectric coefficients and the measured piezoresponse, however the quantity d_{zz} has a complex dependence on the piezoelectric coefficients and the geometry of the applied electric field and sample crystal structure. For a tetragonal ferroelectric $d_{zz}(\theta) = (d_{31} + d_{15}) \sin^2 \theta \cos \theta + d_{33} \cos^3 \theta$ [10]. Where d_{31}, d_{15} and d_{33} are the materials piezoelectric coefficients and θ the angle between the applied electric and the polar c-axis. Thus the quantification is dependent on a knowledge of several of the piezoelectric coefficients (which may be different locally, compared to the values measured at the macroscale) and the angle (θ) the electric field makes with respect to, in the case of a tetragonal crystal, the polar c-axis. Vertical PFM measures a scalar response, which means it cannot distinguish between a large intrinsic piezoelectric response of a crystal where the polar axis makes a large angle to the applied field compared to a small intrinsic piezoelectric response at a small angle to the applied field. It is also unable to unambiguously identify 90 ° domain boundaries (a common domain configuration in tetragonal ferroelectrics). PFM is only able to identify particular domain configuration through geometrical relationships relying on prior knowledge of the crystal structure. There are also numerous issues involving the measurement techniques itself these include, but are not limited to: the effects of electric field divergence and topography on the resolution, sub-surface contributions to the response, noise contributions to the background response, non-linear material response, the effects of the electric field on the state of the material [11] and adsorbed surface layers [12].

Quantified measurements of piezoelectric properties have been carried out predominantly on single crystals [13,14] with some work on thin films [15-17]. These experiments have relied on measuring the crystallographic orientation of a sample using macroscopic techniques. X-ray diffraction can be used to identify a certain crystallographic orientation, which is then preserved with special cuts of the crystal and mounting procedures. Macroscopic techniques are only suitable for materials where the atomic arrangement is preserved throughout the sample. Where the crystal structure of the sample changes on a small length scale these techniques are no longer applicable. We have shown in this paper how this problem can be addressed using electron backscatter diffraction (EBSD) to identify the crystallographic orientation with nanoscale resolution.

EBSD is a scanning electron microscope (SEM) based technique that records the diffraction patterns (called Kikuchi patterns) generated by Bragg reflection of incident electrons by crystallographic planes of the sample. EBSD measurements are attained through a specific geometric setup (including tilting the sample surface to ~70 °, 20 ° to the incident electron beam) and the use of a phosphor screen and a lowlight camera. The Kikuchi patterns are defined by the crystal structure of the sample within a thin surface layer (typically <100 nm). Further details can be found in [18,19].

In this work EBSD can be used to investigate uncertainties in the vertical PFM response by identifying the crystallographic axes of discrete crystallographic regions. EBSD alone can identify the direction, but not the polarity of the polar axis [20]. There are a limited number of studies that have combined PFM with EBSD [21-24]. In most of these resolution was too coarse for the detailed study of nanoscale domain structures. Difficulties relating to the insulating nature of ferroelectric materials and deposition of a carbon layer under the electron beam have until recently limited the achievable resolution of EBSD. Advances such as the optimisation of beam conditions and Kikuchi-pattern recording parameters have allowed for the application of EBSD technique to ferroelectric materials whilst achieving much higher resolution. This has recently been demonstrated for ferroelectric domains [26] but these results were not correlated with PFM.

In this paper we report combined PFM and EBSD measurements of nanoscale domain structures in polycrystalline $Pb(Zr_{0.4}Ti_{0.6})O_3$ (PZT), the dominant piezoelectric material in use today. A geometrical model is used to predict the piezoresponse from the crystallographic data obtained by EBSD, and this is compared to the piezoresponse measured directly with PFM and quantified through two different methods.

Experimental Procedure

A polycrystalline $Pb(Zr_{0.4}Ti_{0.6})O_3$ sample was prepared using conventional powder processing routes. The sample was prepared for microscopy by sectioning and polishing after [25]. Slight relief of the grain boundaries and domain structure was achieved by etching the sample in a dilute mixture of HCl and HF. The base of the sample was sputtered with gold to provide an electrical contact. For SEM based analysis the sample was attached directly to an aluminum SEM stub using silver-loaded epoxy resin. No carbon coating was applied.

For the PFM measurements the sample was attached to a base that provided electrical isolation from the PFM apparatus with the base of the sample connected to electrical ground separately.

PFM was carried out on a Bruker Icon AFM with in-built lock-in amplifier. An in-built signal generator was also used to apply the oscillating voltage to the sample. Electrically conductive tips coated with Pt/Ir (Nanosensors PPP-EFM) were used. The cantilevers were chosen with a relatively high force constant (4.3 N/m) to minimize electrostatic effects [26]. The sensitivity of the AFM was recorded using force curves thus converting measurements in volts to a value in meters. The PFM measurements were all recorded with a driving voltage of 3 V at a frequency of 15 kHz. This frequency was chosen to minimize the background resonances of the experimental setup. A calibration of the PFM measurements was made closely following the method outlined in [27].

EBSD maps were recorded on a Zeiss Supra 40 field emission SEM using Oxford Instruments Channel 5 EBSD software. The experimental method follows the procedure detailed by Farooq et al [25, 26], with the exception that high current mode with an aperture of 60 µm and an accelerating voltage of 10 kV were used. A recording and indexing rate approaching 10 points per second was made possible through an optimization of the contrast in the Kikuchi patterns. The orientations recorded with EBSD were calibrated by the use of a silicon single crystal reference sample. Further refinement of the EBSD setup was completed with iterative fitting of the sample diffraction patterns giving expected errors in the measured orientations of no more than a few degrees [19]. Maximum angular resolution of the Kikuchi patterns was achieved by bringing the detector as close as possible to the sample. This was an important step in distinguishing 90° domains in the sample.

The geometrical setup of EBSD relies on the tilting of the sample to 70 °. This results in the electron beam forming an elliptical footprint on the sample. This means that the electron beam impinging on the sample is approximately three times larger parallel to the tilt direction compared to its perpendicular. As a result domains with their length running parallel to the tilt direction were chosen for EBSD measurements.

In order to identify the exact same region in the PFM as that measured with EBSD the area was recorded at a number of magnifications in the SEM. Once the same area was located high-resolution PFM scans were performed.

A mathematical model as described in [22] was used to predict the piezoresponse of the material. This was made possible through knowledge of the nanoscale orientation of the sample as determined by EBSD and the voltage applied in the direction of the surface normal of the sample.

Results and discussion

A secondary electron micrograph of the sample is shown in Fig. 1a, topography provides the contrast, which was developed during etching. It is possible to determine grains and a complex domain structure within each grain. For comparison the same, but smaller, region is shown in Fig 1b which shows the topography recorded with the AFM.

An orientation map recorded with EBSD is shown in Fig 2 where the different colours identify a different crystallographic orientation. A step (i.e. pixel) size of 25 nm was used to record the map. The wide blue-coloured domains and the narrow orange-coloured domains form an angle of ~90 ° to each other. Average angular relationship across the map was measured at 89.4 °. The green pixels are mis-indexed pixels and the black pixels are un-indexed. Mis-indexing occurs due to the pseudo-cubic nature of PZT and the un indexed pixels can occur due to over -

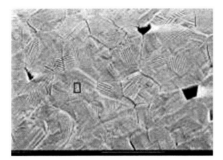 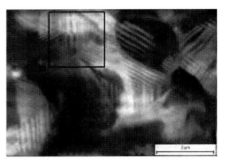

Fig. 1a: Secondary electron image showing general grain and domain structure of PZT ceramic. Black rectangle indicates location of EBSD map.

Fig. 1b: Atomic force microscope image showing surface topography of the same, but smaller, area as Fig. 1b. Location of PFM imaging is indicated.

lapping or poor quality Kikuchi patterns. The EBSD data presented was not subject to any post processing. Whilst the orientation relationship between the blue and orange regions is ~90 ° the polar c-axis is inclined approximately 30 ° to the surface plane for both regions. The average angle to the surface plane for the orange pixels is 28.3 ° and 26.5 ° for the blue pixels.

A PFM image was recorded for the exact same area as the EBSD map as shown in Fig. 3. For comparison an image of the EBSD area is shown to confirm this (Fig 1a and 1b). Fig 3b shows the piezoresponse amplitude (also known as R signal) of the lockin amplifier.

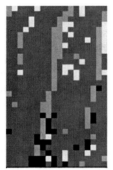

Fig. 2: Electron backscatter diffraction orientation map of the region indicated in Fig. 1a

An area of low contrast (the dark oval in the piezoresponse, Fig 3b) was observed at the exact location where the EBSD map was recorded. This can identified as the brighter region in the topography image (Fig. 3a) and is thought to be a carbon layer, about 5 nm thick, deposited under the electron beam during the EBSD experiment. The resolution is suggested to be at least 30 nm from the smallest feature size recorded.

When the EBSD image (Fig. 2) is compared to Fig. 3 it can be seen that the domains imaged with EBSD, which have been left under a carbon layer, continue outside this region. Because of the carbon layer the piezoresponse of the individual domains directly within the EBSD region have reduced piezoresponse but the piezoresponse of these same domains immediately adjacent to the EBSD region can be measured. Despite the very small difference in the out of plane component of the piezoresponse (28.3 ° compared to 26.5 °) it is possible to see a difference in the amplitude of the response. The orange pixels (28.3 ° to the surface plane) have a larger response as expected as the polar axis is inclined further out of plane.

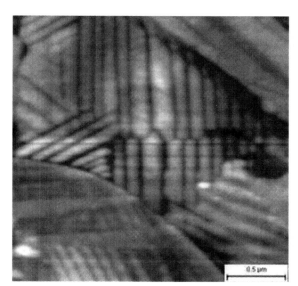

Fig. 3a: Atomic force microscope image showing the topography for the location where the PFM results were obtained. Full Z range is 42 nm.

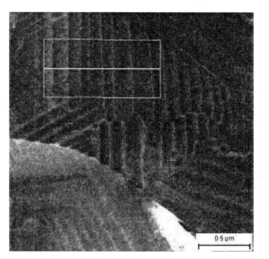

Fig. 3b: Piezoresponse amplitude image for the location indicated in Fig 1. Region of line profile trace (shown in Fig. 4) is indicated.

The measured amplitude of the orange domains is 5.7 mV as seen from the line profile in Fig. 4. The response of the blue domains is found to average at 4.8 mV. Both of these values being measured just outside of the carbon region but within the same set of domains. This value was quantified by comparing the response to the response of a sample with well-known piezoelectric coefficients. A sample of periodically poled LiNbO$_3$ (PPLN) was measured using the same settings before and after imaging the sample under investigation. PPLN has a d$_{33}$ of 7 pm/V and it's domain structure compromises of antiparallel domains so that the difference in response between adjacent domains is 14 pm/V. The response of the sample was also quantified through the calibration of the microscope, although using a comparative sample has potential advantages as it will take account of many of the effects, which affect PFM measurement technique itself, for example the non-uniformity of the field under the tip. Comparing the results to PPLN allows the calculation of the piezoresponse of the orange and blue domains. The response of the orange domains was 26.6 pm/V and for the blue domains 22.4 pm/V. The difference in piezoresponse between the two domain orientations was 4.2 pm/V.

The piezoresponse was also calculated by converting the piezoresponse measured in mV to nm by determining the sensitivity as measured through force distance curves. The results calculated through this method of quantification give a value of piezoresponse of 33.25 pm/V and 28 pm/V for the orange and blue domains respectively. The difference was 5.25 pm/V.

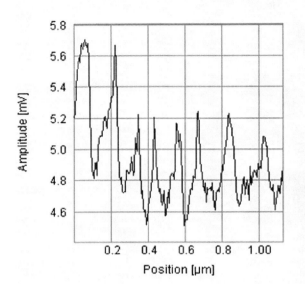

Fig. 4: Piezoresponse amplitude plotted for the line profile shown in Fig 3b.

For further comparison the crystallographic orientations recorded with EBSD were input into the model described earlier and the piezoresponse amplitude predicted. It should be noted that the model describes a monodomain single crystal and the input values for the piezoelectric coefficients are themselves obtained through theory [28]. The applied voltage is also described as a uniform field.

The piezoresponse amplitude predicted for the orange domains was 57.4 pm/V and for the blue domains 53.5 pm/V with a difference of 4.1 pm/V.

The values predicted by the model are higher than those measured directly with PFM. Comparing the measured values of 26.6 pm/V and 33.25 pm/V to the predicted value of 57.4 pm/V for the orange domains and 22.4 pm/V and 28 pm/V compared to 53.5 pm/V for the respective measured and predicted blue domains. Despite this discrepancy the difference between the two domains is in excellent agreement for both the measured and predicted values, at 4.3 pm/V and 5.25 pm/V for the measured values and 4.1 pm/V for the predicted result.

It is likely that the high predicted value is related to the simplicity of the model as described earlier. The level of agreement between the two different methods of quantification is maybe surprising given the fact the values quantified through sensitivity measurements do not take into account any of the complexity related to the measurement technique. The technique simply quantifies the sample surface deflection and assumes the voltage reaching the sample is the same as that applied by the signal generator.

The difference between the measured and predicted values could come from many different sources. These include: in-plane components and variation of the electric

A possible route to the quantification of piezoresponse force microscopy

field through the interaction volume, the effects of electric field divergence and topography on the resolution, sub-surface contributions to the response, noise contributions to the background response, non-linear material response, and the effects of the electric field on the state of the material [11]. It is also well documented that adsorbed surface layers, which are dielectric, will affect the piezoresponse measured by PFM [12]. Such effects are difficult to avoid when operating in ambient conditions when the sample is likely to be exposed to water vapor [31]. Additional surface effects are likely and probably exacerbated when the sample has been subjected to EBSD measurements. Adsorbed layers, as well as affecting the piezoresponse of the material, can also contaminate the scanning tip and affect results. Adsorbed layers could therefore account for some of the discrepancy in the measured and predicted piezoresponse. All of these issues are recognized as important influences on the measured piezoresponse and whilst further discussion of these effects is beyond the scope of this paper, these are active areas of research.

When considering the two methods of quantification there are likely to be differences related to the measurement made on the single crystal PPLN sample and the polycrystalline PZT sample. Whilst both samples experience clamping due to the PFM technique exciting only a local region of the crystal surrounded by the unexcited bulk, the effect is likely to be different between the two samples. When considering the interaction volume excited by the field under the AFM tip there will be no change in the crystal structure in this region for the single crystal. In the polycrystalline sample (and in contrast to the model) there will almost certainly be a range of discrete crystallographic regions that are probed by the applied field. It is currently understood that the PFM is dominated by the volume close (within several tip radii of the tip [14]) to the tip but may extend to nearly 2 microns depth [33]. In a polycrystalline sample this may include regions that enhance or temper the piezoresponse measured at the sample surface by the tip.

It is difficult to judge which of the values is most valid in terms of the true response of the material under applied electric field. As discussed there are many uncertainties in the measurement and effects that are likely to change between experiments and samples. It is certain that the model values presented here are artificially high and the measured values are more likely to have a reduced response if subject to most of undeterminable effects, such as adsorbed surface layers.

It is very encouraging that the quantified values are in such close agreement and that these are well within a factor of three of the predicted values. This result not only ratifies current understanding of PFM, it also confirms the relevance of the two quantification methods presented here.

Conclusions

Whilst it may seem a long way from many fundamental standards the development of advanced standards remains imperative. Standards can be used to ensure the veracity of results and provide a confidence in quoted values. This confidence is almost a prerequisite to the uptake by industry of devices or materials that describe a given performance. In this work the correlation of EBSD and PFM has been used to provide a quantitative measurement of the piezoelectric properties of a sample where the crystallography at the nanoscale is unknown.

This coupled measurement technique links the piezoresponse to the crystallography of the sample at the same scale. We report quantitative measurements of the local piezoresponse of nano-structured ferroelectric materials with a spatial resolution down to 25 nm, commensurate with the size of the domain structure in this material.

The importance of establishing a standard for the quantification of PFM measurements has gained importance as the technique asserts itself as key tool in the characterization of ferroelectric and piezoelectric materials and devices.

Acknowledgements

Thanks to Dr T P Comyn and Prof. A J Bell at the Institute for Materials Research, University of Leeds for the provision of the PZT sample. Thanks also to Dr Ken Mingard of NPL for the very useful discussions surrounding EBSD. This work was supported by The UK's National Measurement Office.

References

1] Guthner P., Dransfield K., Appl. Phys. Lett, **61**, 9, 1137 (1992).

2] Chu Y. H et al, Appl. Phys. Lett, **90**, 252906 (2007)

3] Nath R. et al, Appl. Phys. Lett, **93**, 072905 (2008)

4] Kalinin, S. V., Rar, A., Jesse, S., *A Decade of Piezoresponse Force Microscopy: Progress, Challenges and Opportunities.* Arxiv preprint cond-mat/0509009, (2005)

5] Kholkin A., Kalinin S. V., Roelofs A., Gruverman A., *Scanning probe microscopy: electrical and electromechanical phenomena at the nanoscale*, Springer Science + Business Media, New York (2007)

6] Alexe M., Gruverman A. (Eds), *Nanoscale characterisation of ferroelectric material-Scanning probe microscopy approach*, Springer, Berlin (2004)

7] Kan Y., Lu X., Wu X., Zhu J., App. Phys. Lett, **89**, 262907 (2006)

8] Kim Y., Alexe M., Salje E., Appl. Phys. Lett. **96**, 032904 (2010)

9] Jungk T., Hoffman A., Soergel E., Appl. Phys. Lett. **89**, 163507 (2006)

10] Harnagea C., Pignolet A., Alexe M., Hesse D., Integrated Ferroelectrics, **44**, pp. 113-124 (2002)

11] Gruverman, A., Kalinin, S. V., J. Materials Science **41**, 107-116 (2006)

12] Peter F., Rüdiger A., Dittman R., Waser R., Szot K., Reichenberg B., Prume K., Appl. Phys. Lett. **87**, 082901 (2005)

13] Felten, F., Schneider, G. A., Munoz Saldana, J., Kalinin, S. V. J. Appl. Phys. **96** (1), 563-568 (2004)

14] Tian L., Vasudevarao A., Morozovska A. N., Eliseev E. A., Kalinin S. V., Gopalan V., J. Appl. Phys. **104**, 074110 (2008)

15] Kalinin, S. V., Gruverman, A., Bonnell, D. A., Appl. Phys. Lett. **85** (5), 795-797 (2004)

16] Lin, H-N., Chen, S-H., Ho, S-T., Chen, P-R., Lin, I-N., J. Vac. Sci. Technol. B **21** (2), 916-918 (2003)

17] Wu, A, Vilarinho, P. M., Shvartsman, V. V., Suchaneck, G, Kholkin, A. L., Nanotechnology **16**, 2587-2595 (2005)

18] Randle V., Engler O., *Texture Analysis, Macrotexture, Microtexture and Orientation Mapping*, Gordon and Breach science publishers, (2000)

19] Humphreys F. J., J. Materials Science, **36**, 3833-3854, (2001)

20] Burnett, T. L., Comyn, T. P., Merson E., Bell, A. J., Mingard, K., Hegarty, T., Cain, M. G., IEEE Transactions on Ultrasonics Ferroelectrics and Frequency Control **55** (5) 957-962, (2008)

21] Gupta P., Jain H., Williams D. B., Kalinin S. V., Shin J., Jesse, S., Baddorf, A. P., Appl. Phys. Lett. **87**, 172903, (2005)

22] Yang B., Park, N. J., Seo B. I., Oh, Y. H., Kim, S. J., Hong, S. K., Lee, S. S., Park, Y. J., Appl. Phys. Lett. **87**, 062902, (2005)

23] Lowe M., Hegarty T., Mingard K., Li J., Cain M., Journal of Physics: Conference Series, **126**, 012011, (2008)

24] Garcia R. E., Huey B. D., Blendell J. E., J. of Appl. Phys. **100**, 064105, (2006)

25] Farooq M. U., Villaurrutia R., MacLaren I., Kungl H., Hoffman M. J., Fundenberger J. J., Bouzy E., J. of Microscopy, **230**, 445-454, (2008)

26] Farooq, M. U., Villaurrutia, R., MacLaren, I., Burnett, T. L., Comyn, T. P., Bell A. J., Kungl, H., Hoffmann, M. J., J. Appl. Phys., 104, 024111 (2008)

26] Johann F., Hoffman A., Soergel E., Phys. Rev. B, 81, 094109 (2010)

27] Jungk, T., Hoffmann A., Soergel, E. Appl. Phys. Lett. **91**, 253511 (2007)

28] Haun, M. J., Furman, E., Halemane, T. R., Cross, L. E., Ferroelectrics, **99**, 63-86, (1989)

31] Shin J et al, Nano Lett. **9**, 3720-3725 (2009)

33] Johann, F., Ying, Y. J., Jungk, T., Hoffmann, A., Sones, C. L., Eason, R. W., Mailis, S., Soergel, E., Appl. Phys. Lett. **94**, 172904, (2009)

High resolution analysis of tungsten doped amorphous carbon thin films

Marcin Rasinski[1a, 2], Martin Balden[2], Stefan Jong[2], Philipp-Andre Sauter[2], Malgorzata Lewandowska[1], Krzysztof Jan Kurzydlowski[1],

[1] *Affiliation: Warsaw University of Technology*
Address: Faculty of Material Science and Engineering, Woloska 141, 02-507, Poland
Telephone number: 0048-22-2348724
[a] *mrasin@o2.pl*

[2] *Affiliation: Max-Planck-Institut für Plasmaphysik*
Address: Boltzmannstraße 2, D-85748 Garching, Germany
Telephone number: 0049-89-3299-01

Abstract

Thin W-doped amorphous carbon films were produced by magnetron sputtering at room temperature. The influence of a postannealing treatment on the size, crystallographic structure and distribution of tungsten carbide particles was analyzed. Since the crystallites size after annealing is in the nanometer range, the investigations required the use of High Resolution Scanning Transmission Electron Microscopy (HR STEM). X-ray diffraction (XRD) was performed as complementary technique for crystalline phase analysis. Local nanodiffraction technique was used to address the phase to single crystallites. All three tungsten carbide phases (WC, W_2C, WC_{1-x}) were found depending on W concentration and annealing temperature, sometimes even in the same film.

Introduction

Nuclear fusion is becoming one of most promising renewable power source for the 21 century. The first wall of the next-step fusion reactor ITER (International Thermonuclear Experimental Reactor) will consist of beryllium, tungsten and carbon [1]. Erosion of these wall materials during operation leads to the formation of metal-containing hydrocarbon layers deposited in the plasma vessel [2]. If tritium is used as fuel, this is of safety concern due to the high radioactive inventory [3]. Therefore, it is important to understand the erosion and hydrogen retention properties of such layers, which exhibit far different properties in comparison to the substrates made of carbon, Be and/or W. Metal doped carbon films are used in this context as a model system to study the influence of metals on the erosion of carbon. Such films are studied [4 5] not only due the importance for fusion, but also because of their special mechanical [6], optical, electrical [7] properties.

Different types of tungsten carbides have been reported for the W doped carbon films subjected to annealing [8 9]. However, these carbides have been identified by "global" diffractometry, which not always provide description of their size/shape and spatial distribution. When the size of crystallites are in nanometer scale and the presence of more than one phase is expected (as in the case of tungsten which forms hexagonal WC, W_2C and cubic WC_{1-x} carbides), "local investigations" approach is necessary. Such a possibility offer modern STEM microscopes. In the present work, such a microscope was used to investigate the influence of W concentration and annealing temperature on the structure, size and spatial distribution of tungsten carbides was investigated employing high resolution imaging together with nanodiffraction technique.

Experimental

Amorphous carbon films doped with different tungsten concentration were produced by magnetron sputtering using dual source (graphite and tungsten) cathode with argon as a sputtering gas. A dc magnetron source was used in the case of graphite target and a rf source for the tungsten target – details of the deposition process are described elsewhere [4]. Pyrolytic graphite was selected as substrate. Atomic tungsten concentration in deposited films was 9%, 18% and 22%. Such prepared samples were subjected to heat treatment. Annealing was performed in He atmosphere at 1450, 2200 and 2800 K for one hour. Furnace chamber with samples was purged with He for three times at room temperature and three times at 400°C to remove air and avoid tungsten oxidation during annealing.

The samples after annealing were investigated by X-ray diffraction with Cu K_α radiation. More information on XRD setup is given elsewhere [5]. Since the precipitates, especially at lower annealing temperature, are in nanometer scale, XRD results were ambiguous and more detailed investigation was required. From each sample thin lamella was prepared using single beam Focused Ion Beam (FIB) operating at 40 kV with maximum beam current ~42 nA. Final thinning was performed with 10 kV and beam current below 0.1 nA.

Thin lamellas of the annealed films have been investigated with dedicated STEM equipped with Cs corrector by CEOS, Schottky field emission electron source operated at 200 kV and with beam current ~100 nA. Atomic resolution observations were performed in Bright Filed mode (BF) and High Angle Annular Dark Field (HAADF), also called Z-contrast (contrast generated by electrons which are not Bragg scattered and their intensity is proportional to Z^2). Phase detection for single carbides was performed by nanodiffraction which is a novel and elegant technique employed in a STEM [10] and similar to Convergent Beam Electron Diffraction (CBED) in classical TEM. It provides possibility to obtain diffraction from nanometer scale crystals due to very small beam diameter (2 nm and less, depending on device).

Results and discussion

Fig. 1 presents XRD data of three samples with different tungsten concentrations after annealing at 1450 K for one hour. For the lowest tungsten concentration no carbide phase identification is possible. Broad peaks at around 37°, 63° and 75°

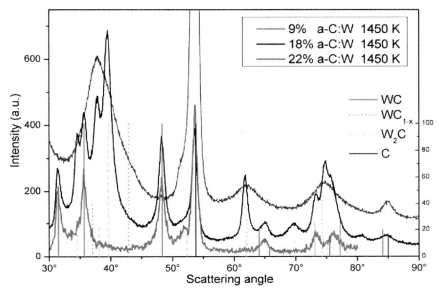

Fig. 1: X-ray diffractograms of W doped films with indicated W concentration annealed at 1450 K. Diffraction peaks for the carbide phases WC, W_2C, WC_{1-x} (PDF card 25-1047, 35-776, 20-1316) and for pure graphite (substrate) are indicated [12].

indicate the presence of very small particles. However, they cannot be clearly attributed to W_2C, WC_{1-x} or mixture of both. For the sample doped with 18% W and annealed at 1450 K more narrow peaks have been recorded. Part of these peaks can be attributed to specific phases, but some part of the spectrum is still ambiguous. In the range from 30 to 50° identification of WC and W_2C phases is possible. Nevertheless peaks around 63° and in the range 70-80° remain uncertain. Most firm conclusions can be made with regard to the sample with 22% of W and annealed at 1450 K. Relatively narrow peaks of WC phase are visible. For all samples the peak at 54° can be attributed to graphite substrate. Additional XRD results can be found elsewhere [11].

Electron microscopy provides more information about type, size and distribution of the carbide crystallites after annealing. The structure of film 9% W doped and annealed at 1450 K is very homogenous and contains carbides typically 3-6 nm in diameter, as shown in Fig. 2. Out of 23 analyzed diffraction patterns taken from individual particles 17 were cubic WC_{1-x} and 6 hexagonal W_2C, as presented in Tab. 1. Based on these results, the ambiguous broad XRD peaks can be identified as caused by a mixture of nanometer scale WC_{1-x} and W_2C crystallites.

STEM observations show that the annealing of 9% W doped film at 2200 K leads to the formation of 10 nm size carbides as shown in high resolution images in Fig. 3. cd. The predominant phase is WC1-x carbide with the presence of some W$_2$C carbides.

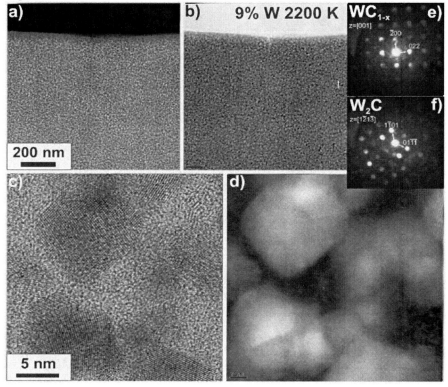

Fig. 2. STEM images in BF mode (a,c) and in Z-contrast (b,d) together with typical electron diffractograms of WC$_{1-x}$ (e) and W$_2$C (f) of 9% W film annealed at 1450K

Significant structural transformation is observed in 9% W film annealed at 2800 K. As presented in Fig. 4, carbide size diversity can be noted. In addition to the carbides in the range of tens of nm, a small number of larger carbides, up 1μm and more, are detected. Nanodiffraction analysis reveals that more than half of the analyzed crystallites are WC1-x but W2C carbides are no longer present. Instead hexagonal WC carbides are formed. Large carbides, of the size 1 μm and more, are found to be exclusively WC (right side of Fig. 4 ab). However, WC carbides are also present as small crystallites.

Carbide structure in the film doped with 18% W and annealed at 1450 K is shown in Fig. 5. Carbides are larger and more densely arranged in comparison to 9% W film annealed at the same temperature. Already at 1450 K all three types of carbides are present with W$_2$C being dominant. Out of 28 analyzed carbides only 3 were WC and 7 WC$_{1-x}$

High resolution analysis of tungsten doped amorphous carbon thin films 111

Tab. 1 Amount of identified carbides depending on initial W concentration and annealing temperature

	WC_{1-x}	W_2C	WC
9% W 1450 K	17	6	
9% W 2200 K	11	9	-
9% W 2800 K	13	-	9
18% W 1450 K	7	18	3
18% W 2200 K	4	11	15
22% W 1450 K	-	-	24

Annealing at 2200 K changes the proportion of carbides. The predominant phase is WC, however, size of carbides becomes inhomogeneous, as presented in Fig. 6. The highest density of carbides is observed in the film doped with 22% W and annealed at 1450 K (Fig. 7). Only WC phase was identified (24 diffractograms analyzed) in this sample, which is in good agreement with XDR data.

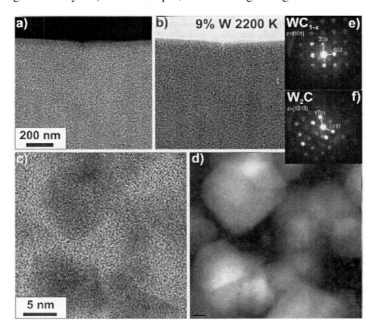

Fig. 3: STEM images in BF mode (a,c) and in Z-contrast (b,d) together with typical electron diffractograms of WC_{1-x} (e) and W_2C (f) of 9% W film annealed at 2200K

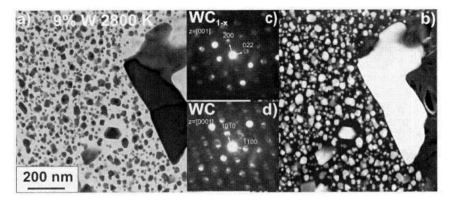

Fig. 4: STEM images in BF mode (a) and in Z-contrast (b) together with typical electron diffractograms of WC_{1-x} (c) and WC (d) of 9% W film annealed at 2800

The results of STEM investigation show that the size and carbide density is determined by the annealing temperature and W concentration. The higher W concentration the larger and more dense are carbides. The results suggest that the density of WC_{1-x} does not change much during annealing and depends inversely on the W concentration. In the sample doped with 9% W, the amount of WC_{1-x} is the highest, while at 22% W, WC_{1-x} phase was not detected at all.

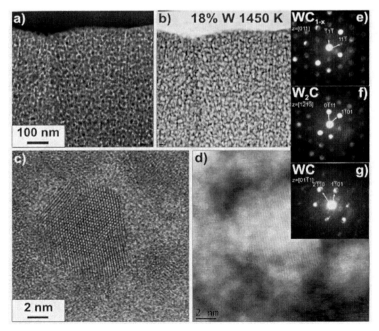

Fig. 5: STEM images in BF mode (a,c) and in Z-contrast (b,d) together with typical electron diffractograms of WC_{1-x} (e) W_2C (f) and WC (g) of 18% W film annealed at 1450K

High resolution analysis of tungsten doped amorphous carbon thin films 113

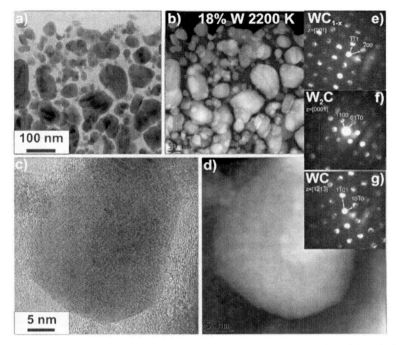

Fig. 6: STEM images in BF mode (a,c) and in Z-contrast (b,d) together with typical electron diffractograms of WC_{1-x} (e) W_2C (f) and WC (g) of 18% W film annealed at 2200K

The annealing temperature influences the amount of WC and W_2C. In the case of 9% W sample, WC phase starts to appear between 2200 and 2800 K (additional temperature step is required). On the other hand, in 18% W and 22% W samples, WC is detected already after annealing at 1450 K. This means that the precipitation temperature must be lower.

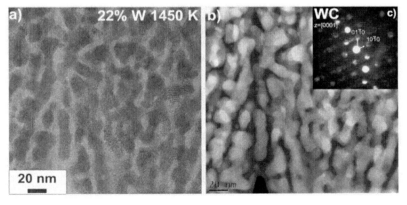

Fig. 7: STEM images in BF mode (a) and in Z-contrast (b) together with typical electron diffractograms WC (c) of 22% W film annealed at 1450K

The W_2C carbides are present in 9% W and 18% W samples annealed at 2200 K while the sample doped with 22% of W exhibits no W_2C even at 1450 K. Lower annealing temperature of 22% W sample is required to check whether W_2C is formed and what is the lowest precipitation temperature for it. In the case of 18% W sample, lower annealing temperature is necessary to check precipitation of WC carbides.

Conclusion

The effect of tungsten concentration and annealing temperature on tungsten carbide crystallite formation in amorphous tungsten doped carbon films was investigated. X-ray diffraction gives hints and general information on carbide phase type but often is ambiguous, especially with low tungsten concentration. Broad peaks do not give clear results.

STEM imaging performed in the present study provides insight into tungsten carbide formation in amorphous tungsten doped carbon films. In the 9% W doped film annealed at 1450 and 2200 K, WC_{1-x} as well as W_2C carbides were detected. Annealing at 2800 K resulted in the presence of WC_{1-x} and WC carbides. In the case of 18% W concentration, annealing at 1450 and 2200 K leads to precipitation of all three carbides with the domination of W_2C at 1450 K and WC at 2200 K. For a sample containing 22% W, only WC phase precipitation was observed even after annealing at 1450 K.

Acknowledgments

The study has been partly financed by the statutory funding of the Materials Design Group of the Faculty of Materials Science and Engineering at Warsaw University of Technology.

References

1. Holtkamp N et al (2009) The status of the ITER design. Fusion Eng. and Des. 84 (2-6):98-105
2. Roth J et al (2009) Recent analysis of key plasma wall interactions issues for ITER. J. of Nucl. Mater. 390-91:1-9
3. Causey RA et al (2002) Tritium inventory and recovery in next-step fusion devices. Fusion Eng. and Des. 84 61-62:525-536
4. Balden M Characterization of nano-structured W, Ti, V and Zr doped carbon films. Thin Solid Films (to be published / submitted)
5. Adelhelm C et al Investigation of metal distribution and carbide crystallite formation in metal-doped carbon films (a-C:Me, Me=Ti, V, Zr, W). Surf. and Coat. Technol. (to be published / submitted)

6. Ming X. et al (2009) Mechanical properties of tungsten doped amorphous hydrogenated carbon films prepared by tungsten plasma immersion ion implantation. Surf. and Coat. Technol. 203 (17-18):2612-2616
7. Abad MD et al (2009) WC/a-C nanocomposite thin films: Optical and electrical properties. J. of App. Phys. 105(3):105-110
8. Mrabet SE et al (2009) Thermal Evolution of WC/C Nanostructured Coatings by Raman and In Situ XRD Analysis. Plasma Process. Polym.6:444–449
9. Kurlov AS and Gusev AI (2007) Neutron and x-ray diffraction study and symmetry analysis of phase transformations in lower tungsten carbide. Phys. Rev. B 76 174115
10. Cowley JM (2004) Applications of electron nano diffraction J. M. Cowley. Micron 35 (5):345-36011.Sauter PA et al Carbide formation in tungsten-containing amorphous carbon films by annealing. Thin Solid Films (to be published / submitted)
12. Joined committee for powder diffraction studies - international centre for diffraction data (jcpdsicdd), powder diffraction file, release 2000

Electron Microscopy Studies on Oxide Dispersion Strengthened Steels

Arup Dasgupta, R Divakar, Pradyumna Kumar Parida, Chanchal Ghosh, S Saroja, E Mohandas, M Vijayalakshmi, T Jayakumar and Baldev Raj

Indira Gandhi Centre for Atomic Research,
Kalpakkam – 603 102, INDIA

Abstract The 9Cr ODS steel, a candidate material for fast fission and fusion reactor applications, derives its superior irradiation performance due to the dispersoids in ferrite matrix. Electron microscopy studies on mechanically alloyed Fe-Y_2O_3-Ti model alloys and 9Cr yttria/Ti dispersion strengthened (ODS) ferritic steels are discussed in this paper. The size distribution of dispersoid in the consolidated model alloy and the ODS steel were found to be peaking at ~ 15 nm and 5 nm, respectively. The porosity in the ODS steel was greatly reduced in comparison to the model alloy owing to a superior milling process. The dispersoids were identified as Y-Ti-O complexes. An orientation relationship between the yttria dispersoid and the ferrite grains, in which they are embedded, was observed in the model alloy. Microtexture analysis on sections of consolidated ODS alloy rods showed a [110] fiber texture typical of rods of bcc metals. The ODS steel tube however contained randomly oriented ferrite grains.

1. Introduction

The oxide dispersion strengthened (ODS) steels are candidate core structural materials for both fast fission as well as fusion reactors. The core structural materials, namely the clad and wrapper in case of fission reactors, have to withstand high temperatures and intense neutron irradiation. In sodium cooled fast reactors, these components are required to be stable upto a temperature of 973 K over prolonged periods of time in addition to achieving a fuel burn up of about 200 GWd/t. Such high burn-up is required for deriving economic benefits. Neutron flux to which the core materials are subjected to is of the order of 10^{19} n $m^{-2}s^{-1}$ [1]. The materials must resist void swelling and irradiation creep in such hostile environment [2-3]. The 9-12Cr ferritic ODS steels are an excellent choice from these points of view.

The ferritic ODS steels have superior void swelling resistance by virtue of the inherently favorable interaction between point defects. Besides, the dispersoids also act as trapping sites for irradiation induced point defects thereby enhancing the swelling resistance. In addition, the dispersed oxide particles impede the movement of dislocations providing the required high temperature creep resistance, which is otherwise low in ferritic steels.

In the last two decades, there have been considerable research efforts on development of ODS steels [4-9]. These research efforts encompass synthesis of the alloy by powder metallurgical processing methods such as mechanical alloying followed by consolidation employing hipping, forging or extrusion at high temperatures. The 9Cr ODS clad tubes have been reported to possess isotropic properties, while the 12 Cr ODS was reported to be more anisotropic [10-12]. Moreover, the former have shown lower DBTT shift on irradiation and superior formability [13]. The addition of a small amount of Ti has been reported to be responsible for refinement of the dispersoid size [14-17] through the formation of Y-Ti complex oxide, which in turn significantly improves creep rupture strength. It has been reported in the literature [18-20] that these Y-Ti complex oxide particles are exceptionally stable under irradiation environments.

This paper presents the results of an electron microscopy investigation on the microstructural, microchemical and micro-texture characteristics of the Fe-0.2Ti-0.35Y_2O_3 model alloy and Fe-9Cr-2W-0.2Ti-0.35Y_2O_3-0.11C steel.

2. Experimental

Two alloys, namely Fe-0.2Ti-0.35Y_2O_3 (called as model ODS alloy) and Fe-9Cr-2W-0.2Ti-0.35Y_2O_3-0.11C (called as ODS steel), have been studied using microscopy techniques. These systems have been characterized in the powder as well as in consolidated form. Consolidation was achieved by hot forging in case of the model ODS alloy and hot extrusion for the steel, at 1323 K. SEM analysis was carried out in a FEI ESEM model XL-30 operated at 30 kV in secondary electron and back-scattered electron imaging modes for morphological and topological characterization, energy dispersive X-ray spectroscopy (EDS) for microchemistry and electron back scatter diffraction (EBSD) for microtexture information. TEM studies were carried out in a FEI CM 200 Analytical Transmission Electron Microscope and a JOEL 2000 EX (II) High Resolution TEM. A few select specimens were studied in FEI Tecnai G2 F30 operated at 300 kV TEM at Indian Institute of Science, Bengaluru, India. The powder specimen for TEM studies was prepared by mixing it thoroughly in a thermal-set, low vapor pressure epoxy, filling in a 3 mm inner diameter Teflon ring and curing at 403K for ~ 60 minutes. This was subsequently ground, polished, dimpled and ion-milled as for conventional TEM samples.

3. Results and Discussion

3.1 Characterization of model ODS alloy

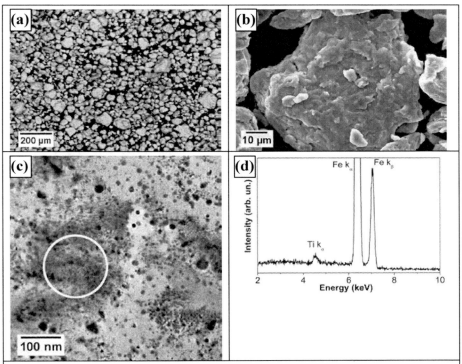

Fig. 1. Micrographs of ball milled powder of the model ODS alloy; (a-b) SEM micrographs, (c) TEM micrograph and (d) EDS spectra from encircled area in (c).

Figure 1(a) shows the secondary electron micrograph of the Fe-0.2Ti-0.35Y_2O_3 powder ball milled for 48 hours, spread on conducting carbon tape. It is observed that the mechanically alloyed powder particles vary in size between 2 – 100 µm, with the peak at 3.5 µm. Since agglomeration can often lead to miscalculation of particle sizes, it is necessary to exercise caution and care in this evaluation procedure. Therefore, the number quoted as particle size is at best representative unless measured separately for each particle, at higher magnification. Figure 1(b) shows the microstructure of an individual particle. It is seen that this milled powder particle measures about 75 µm and does not appear to be an agglomerate. The microstructure is convoluted lamellar in nature indicating incomplete milling. The lamellar structure is characteristic of the mechanical milling procedure, which is a dynamic process of welding and fragmentation of the feed powder particles. Figure 1(c) shows a typical TEM micrograph of a milled powder particle. It is observed that the powder is composed of many smaller particles embedded in a matrix. Figure 1(d) shows the EDS spectra from a region circled in Fig. 1(c). Composition evaluation from Fe and Ti peaks confirm the intended chemical composition of the powder material. However, Y could not be detected due to the low weight fraction (0.35%). Therefore, further milling or a different milling process is required to obtain a well blend lamellar structure.

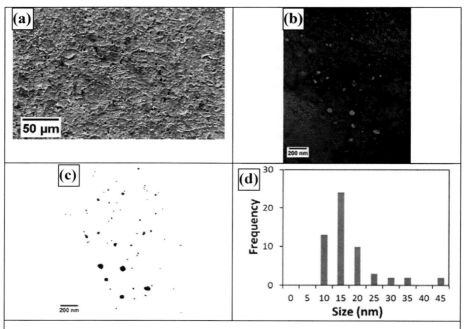

Fig. 2. Micrographs of the ODS model alloy in hot forged condition; (a) SEM micrograph, (b) Dark field TEM micrograph corresponding to (222) of yttria, (c) Processed image of (b) and (d) Histogram of yttria size distribution.

Figure 2(a) shows the secondary electron micrograph of the product consolidated by hot forging at 1323 K. It is observed that a small pore density of <1% is present, which is attributed to the convoluted lamellar structure of the milled powder described in Fig.1. TEM and electron diffraction analysis [21] showed that BCC yttria-titania complex dispersoids were embedded in a ferrite matrix. Diffraction spots from yttria were used to obtain dark field TEM micrographs corresponding to the yttria crystallites to study their size and distribution. Figure 2(b) shows such a dark field TEM micrograph corresponding to (222) of yttria bearing an inter-planar spacing of 0.306 nm. Figure 2(c) shows a processed image derived from Fig. 2(b), obtained by suitable thresholding of the raw image. The figure represents the distribution of the yttria dispersoids in the matrix. Figure 2(d) shows the histogram of yttria size distribution. From these figures it is observed that while the distribution is more or less uniform in the area under examination, the size distribution peaks at 15 nm. The average inter particle distance is measured as 80 nm. Since, dislocation pinning depends critically on the size and inter-particle distance [16-17], it is concluded that further refinement of the dispersoids is required. An optimum particle size and inter-particle separation is of the order of 5 nm [22], and 20 nm (24 nm for 7 nm particles [17]), respectively.

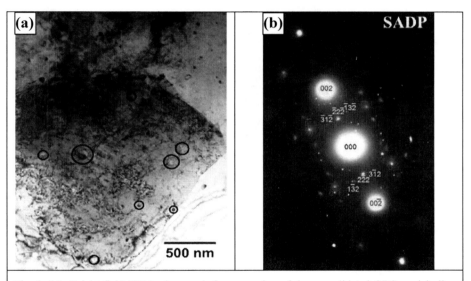

Fig. 3. (a) Bright field TEM micrograph from a region of the consolidated ODS model alloy showing a ferrite grain with a number of dispersoids marked by circles, and (b) SADP from the ferrite grain showing OR between yttria and ferrite.

Figure 3(a) is a bright field TEM image of a ferrite grain in which yttria dispersoids are scattered. Some of them are encircled to indicate their location within the ferrite grain. Figure 3(b) shows the selected area diffraction pattern (SADP) from this ferrite grain. The analysis of the SAD pattern revealed the following orientation relationship between the ferrite and yttria:

$$(002)_{Ferrite} // (222)_{Yttria}, and$$
$$[100]_{Ferrite} // [\bar{1}12]_{Yttria}$$

Eqn. 1

A different kind of OR (Eqn. 2) has earlier been reported for the Eurofer 97-ODS alloy [23].

$$(1\bar{1}0)_{Ferrite} // (1\bar{1}\bar{1})_{Yttria}, and$$
$$[111]_{Ferrite} // [110]_{Yttria}$$

Eqn.2

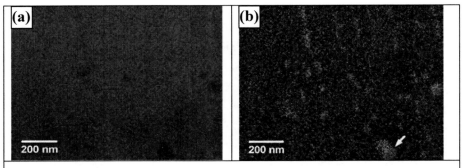

Fig. 4. EDS maps corresponding to, (a) Fe-k$_\alpha$ (blue) alone, and (b) RGB image in which, red, green and blue represents Ti-k$_\alpha$, Y-L$_\alpha$ and Fe-k$_\alpha$, respectively.

Presence of OR between the dispersoid and the matrix suggests precipitation of the dispersoids during the consolidation process. In fact, Zhang et al. [17] have shown through a systematic experiment in Co-base ODS alloys that yttria dissolves during mechanical alloying and re-precipitates at 963 K, beyond which they coarsen. The coarsening can however be arrested by incorporation of elements with high affinity for oxygen, like Ti, which diffuse through the iron matrix to yttria sites faster than yttrium. The Ti attaches itself to the yttria molecule and thereby restricts coarsening of the particle. The affinity of Ti to yttria is conclusively demonstrated in Figure 4. The figures show EDS maps corresponding to, (a) Fe-k$_\alpha$ (blue) alone, and (b) RGB image in which, red, green and blue represent Ti-k$_\alpha$, Y-L$_\alpha$ and Fe-k$_\alpha$, respectively. Fig. 4(a) shows Fe deficient (dark regions) while Fig. 4(b) shows that these regions contain Y-Ti complexes. Although, O cannot be reliably mapped using EDS technique, the phase analysis results presented earlier, indicate that these complexes are of Y-Ti-O type. Formation of Y-Ti-O nanoclusters has been reported in literature [24]. A closer look at the dispersoid (arrow marked in Fig. 4(b)) reveals that there are two chemical phases in this relatively large dispersoid, a Y-Ti-O complex and Ti rich phase. On the other hand, the smaller dispersoids correspond to only Y-Ti-O. Further insight into the chemistry of coarse particles in presence of excess Ti is under study.

3.2. Characterization of 9-Cr ODS steel

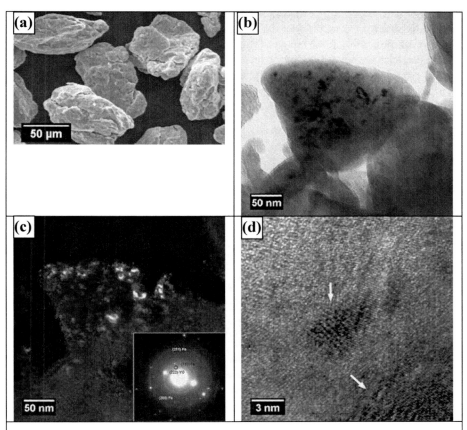

Fig. 5. Micrographs of 9Cr ODS steel powder attritor milled for 4 hours; (a) SEM micrograph (b) and (c) TEM bright field micrograph and dark field micrograph using diffraction from (222) of yttria (the diffraction pattern is shown in the inset), respectively, (d) typical high resolution TEM image of a powder particle (arrows indicate Moiré fringes corresponding to embedded dispersoids).

Figure 5(a) shows the secondary electron micrograph of the ODS steel powder high energy attritor milled for 4 hours. A well blend lamellar structure, devoid of loose particles, with nearly uniform shape and size are observed in contrast to the convoluted lamellar structure (fig. 1(b)). This improvement is attributed to a more efficient milling process. Figures 5(b) and 5(c) show TEM bright field and dark field micrographs, using diffraction from (222) of yttria, respectively, with the corresponding SADP shown as an inset. Diffraction spot corresponding to (222) of yttria (YO) and diffraction rings corresponding to (110), (200) and (211) of the ferrite phase (Fe) are marked in the figure. From Figs. 5 (b) and (c), it is observed that yttria particles with sizes between 5-20 nm are well dispersed in the ferrite matrix. Such a uniformly dispersed matrix in a well blend structure is ideally suited to obtain a dense product during consolidation. Figure 5(d) shows a typical high resolution (HR) TEM image of the yttria dispersoid in the ferrite matrix. The yttria particles are identified by the relatively wider Moiré fringes (indicated by arrows) resulting from the embedded yttria crystallites. The distortion of the lattice fringes indicates high defect density, due to the high deformation during the milling process.

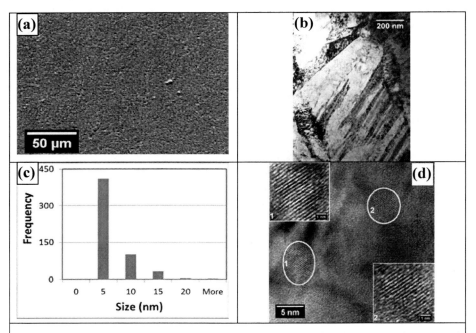

Fig. 6. (a) SEM micrograph of the consolidated ODS steel, (b) Bright field TEM micrograph of this material, (c) Histogram of the yttria size distribution and (d) HRTEM image from a representative region in the alloy showing two yttria dispersoids marked as (1) and (2), lattice fringes from these dispersoids are shown enlarged in the insets.

Figure 6 (a) shows the secondary electron micrograph of the consolidated (hot extruded) ODS steel rod. In contrast to Fig. 2(a), the negligible porosity is attributed to the superior attritor milling process, wherein milling duration is reduced by a factor of 12. Figure 6(b) shows a TEM bright field image of the alloy. A typical martensitic structure with fine laths measuring 50 – 200 nm in width is seen. Dark field imaging corresponding to yttria diffraction was used to evaluate the dispersoid size and distribution, as shown in the histogram in Figure 6(c). It is seen that majority of the dispersoids measure about 5nm. Figure 6(d) shows the HRTEM image from a representative region with two yttria dispersoids marked as (1) and (2) bearing an inter-particle separation of about 15 nm, lattice fringes from these dispersoids are shown enlarged in the insets. The lattice fringes in either dispersoid correspond to 0.29 nm indexed as (222) of yttria. These dispersoids are also expected to be nanoclusters of Y-Ti-O, since the weight percentage of Ti and yttria are identical to that of the model alloy (Fig.4).

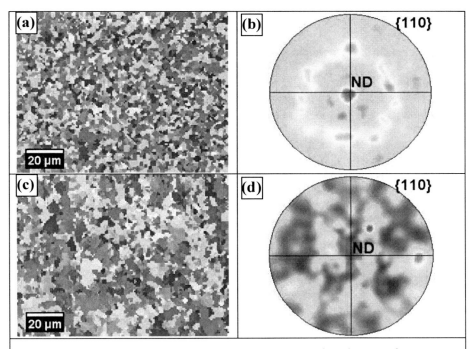

Fig. 7. (a) IPF colored (red: (001), blue: (111) and green: (101)) orientation map of a representative region of the consolidated ODS steel rod, obtained using EBSD, (b) {110} pole figure corresponding (a), (c) IPF colored orientation map of a representative region of a tube fabricated from the consolidated ODS steel rod, (d) {110} pole figure corresponding (c).

Figure 7(a) shows the IPF (Inverse Pole Figure) colored (red: (001), blue: (111) and green: (101) orientation map of a representative region of the extruded rod, obtained using EBSD. Average grain size is measured to be about 3.5 μm. Figure 7(b) shows the {110} pole figure. A concentration of poles (red color) along the normal direction (ND) indicates that [110] directions are predominantly inclined along ND in the extruded rods. This texture is characteristic of a typical [110] fiber texture commonly observed in bcc metals [25]. Figure 7 (c) shows the IPF colored orientation map of a representative region of a tube fabricated from the consolidated rod and Figure 7(d) shows the {110} pole figure of the corresponding region. From Figs. 7(c and d), it is observed that although there is negligible change in grain size, there is no preferential orientation of grains, as is expected in 9Cr steels [10-12]. Such randomly oriented ferrite grains is considered to be a positive attribute since anisotropy in properties is undesirable from application point of view.

4. Conclusions

Two systems, namely a model alloy of Fe-0.2Ti-0.35Y_2O_3 and Fe-9Cr-2W-0.2Ti-0.35Y_2O_3-0.11C, an ODS steel have been studied using electron microscopy techniques.. The microstructures due to different processing methods have been compared. The ODS steel employing attritor milling is found to be superior to the model alloy processed by ball milling. In the model alloy, most of the dispersoids measured about 15 nm while in the steel, significant refinement was achieved (5 nm). The role of titanium in refining the yttria complex has been demonstrated, by confirming the dispersoids to be nanoclusters of Y-Ti-O. The OR between the dispersoid and the ferrite matrix in the model alloy, suggested the re-precipitation of fine particles. While the consolidated ODS steel rods possessed a simple [110]//ND type of fiber texture common to rods of bcc metals, the texture was not present in the tube, which is a positive attribute for the intended application.

Acknowledgments

The authors would like to acknowledge the contributions of International Advanced Research Centre for Powder Metallurgy and New Materials (ARCI), Hyderabad and the Nuclear Fuel Complex, Hyderabad, India.

References

1. Bailly H, Menessier D and Prunier C (1999) The nuclear fuel of pressurized water reactors and fast neutron reactors. Lavoisier Publishing, Paris 16-36.
2. Mannan S L, Chetal S C, Baldev Raj and Bhoje S B (2003) Selection of materials for prototype fast breeder reactor. Proc. of Seminar on "Materials R & D for PFBR" Eds. Mannan S L and Mathew M D, Kalpakkam, India, Jan. 1-2:9-43.
3. Lallement R, Maikailoff H, Mustellier J P, and Villeneuve J (1989) Fast breeder reactor fuel design principles and performance. Nucl. Energy 28:41-49.
4. Huet J J and Leroy V (1974) Dispersion-strengthened ferritic steels as fast reactor structural materials. Nucl. Technol. 24:216-224.

5. Hack G A (1984) Developments in the production of oxide dispersion strengthened superalloys. Powder Metall. 27:73-79.
6. Ukai S, Harada M, Okada H, Inoue M, Nomura S, Shikakura S, Asabe K, Nishida T and Fujiwara M (1993) Alloying design of oxide dispersion strengthened ferritic steel for long life FBRs core materials. J. Nucl. Mater. 204:65-73.
7. Murty K L and Charit I (2008) Structural materials for Gen-IV nuclear reactors: challenges and opportunities. J. Nucl. Mater. 383:189-195.
8. Alinger M J, Odette G R and Hoelzer D T (2009) On the role of alloy composition and processing parameters in nanocluster formation and dispersion strengthening in nanostructured ferritic alloys. Acta Mater. 57:392-406.
9. Verhiest K, Almazouzi A, Wispelaere N De, Petrov R and Claessens S (2009) Development of oxides dispersion strengthened steels for high temperature nuclear reactor applications. J. Nucl. Mater. 385:308-311.
10. Whittenberger J D (1982) Crystallographic texture in oxide dispersion-strengthened alloys. Mater. Sci. and Engg. 54:81-83.
11. Ukai S, Mizuta S, Fujiwara M, Okuda T and Kobayashi T (2002) Development of 9cr-ods martensitic steel claddings for fuel pins by means of ferritic to austenitic phase transformation. J. Nucl. Sci. Technol. 39:778-788.
12. Ukai S, Okuda T, Fujiwara M, Kobayashi T, Mizuta S and Nakashima H, (2002) Characterization of high temperature creep properties in recrystallized 12Cr–ODS ferritic steel claddings. J. Nucl. Sci.Technol. 39:872-879.
13. Ukai S, Kaito T, Ohtsuka S, Narita T and Sakaswgawa S (2006) Development of optimized martensitic 9Cr-ODS steel cladding. Trans. of the ANS Annual Meeting, June 4-8, Reno, USA.
14. Ohtsuka S, Ukai S, Fujiwara M, Kaito T and Narita T (2004) Improvement of 9Cr-ODS martensitic steel properties by controlling excess oxygen and titanium contents. J. Nucl. Mater. 329-333:372-376.
15. Miller M K, Russel K F and Hoelzer D T (2006) Characterization of precipitates in MA/ODS ferritic alloys. J. Nucl. Mater. 351:261-268.
16. Ratti M, Leuvrey D, Mathon M H and Carlan Y De (2009) Influence of titanium on nano-cluster (Y, Ti, O) stability in ODS ferritic materials. J. Nucl. Mater. 386-388:510-543.
17. Zhang L, Ukai S, Hoshimo T, Hayashi S and Qu X (2009) Y_2O_3 evolution and dispersion refinement in Co-base ODS alloys. Acta Mater. 57:3671-3682.
18. Gan J, Allen T R, Birtcher R C, Shutthanandan S and Thevuthasan S (2008) Radiation effects on the microstructure of a 9Cr ODS alloys. J. Mater. 60:24-28.
19. Kishimoto H, Kasada R, Hashitomi O and Kimura A (2009) Stability of Y–Ti complex oxides in Fe–16Cr–0.1Ti ODS ferritic steel before and after heavy-ion irradiation. J. Nucl. Mater. 386-388:533-536.

20. Certain A G, Field K G, Allen T R, Miller M K, Bentley J and Busby J T (2010) Response of nanoclusters in a 9Cr ODS steel to 1 dpa, 525 °C proton irradiation. J. Nucl. Mater. (in press).
21. Saroja S, Dasgupta A, Divakar R, Raju S, Mohandas E, Vijayalakshmi M, Bhanu Sankara Rao K and Baldev Raj (2010) Development and characterization of advanced 9Cr ferritic / martensitic steels for fission and fusion reactors. J. Nucl. Mater. (in press).
22. Kim I, Choi B Y, Kang C Y, Okuda T, Maziasz P and Miyahara K (2003) Effect of Ti and W on the mechanical properties and microstructure of 12% Cr base mechanical-alloyed nano-sized ODS ferritic alloys. ISIJ Int. 43:1640.
23. Klimiankou M, Lindau R and Moslang A (2003) HRTEM study of yttrium oxide particles in ODS steels for fusion reactor application. J. Cryst. Growth 249:381-387.
24. Miller M K, Hoelzer D T, Kenik E A and Russel K F (2004) Nanometer scale precipitation in ferritic MA/ODS alloy MA957. J. Nucl. Mater. 329-333:338-341.
25. Hu H (1974) Texture of metals. Texture 1:233-258.

Fabrication of Probes for In-situ Mapping of Electrocatalytic Activity at the Nanoscale

Andrew J. Wain,* David Cox, Shengqi Zhou and Alan Turnbull

National Physical Laboratory, Hampton Road, Teddington, TW11 0LW, United Kingdom

* andy.wain@npl.co.uk

Abstract

Scanning electrochemical microscopy (SECM) is well established as a powerful means to spatially map the electrochemical activity of materials. By scanning a microelectrode probe over a surface immersed in an electrolyte solution, spatially resolved information about surface activity can be gained through the measurement of local electrochemical processes. The combination of SECM with atomic force microscopy (SECM-AFM) has been identified as an elegant means to take the resolution of such measurements down to the nanoscale. The integration of an addressable electrode onto dual function scanning force probes is advantageous not only because of the small electrode dimensions, but moreover because this permits scanning at constant tip-surface separation using the AFM cantilever optical feedback. Thus, topographical features can be easily measured and deconvoluted from activity variations, and short working distances can be achieved.

In this work we present two novel approaches to the fabrication of probes for SECM-AFM. One method involves the modification of commercially available metallic needle electric force microscopy (EFM) probes to yield high-aspect ratio coated needle probes with an addressable nanodisk electrode at the apex. The second approach involves a focused ion beam (FIB) milling procedure to integrate a platinum electrode into commercial silicon nitride AFM probes. The former approach is found to be more successful and the imaging capability of these probes is demonstrated using a gold-on-silicon patterned substrate.

1 Introduction

Scanning electrochemical microscopy (SECM) is a scanning probe technique in which a piezo-positioned micro- or nano-electrode is used to probe the local topographical features and electrochemical behavior of surfaces.[1,2] In conventional amperometric SECM, the sample of interest is immersed into an electrolyte solution containing a redox mediator, and the potentiostatic electrolysis of that mediator at the tip of the electrode is used as a means to determine either (a) the separation between the tip and surface, or (b) the reactive nature of the surface. The Faradaic current measured at the tip is sensitive to its local environment; in the vicinity of inert surface features, the hindered diffusion of fresh mediator limits the current (negative feedback), whereas conducting or chemically active regions can regenerate the original oxidation state of the mediator, leading to an enhanced current response (positive feedback). Scanning the tip at a fixed height above the surface then allows the mapping of interfacial processes, giving rise to a variety of applications including the study of corrosion and dissolution processes, biological phenomena and electrocatalytic activity.[3-7]

The spatial resolution of SECM is determined by both the dimensions of the sensing probe and the separation between the probe and surface. Probe diameters typically range from 1 μm – 25 μm, with more recent work demonstrating the use of probes in the sub-micron domain.[8,9] There are two major challenges in taking the spatial resolution of this technique down to the nanoscale, namely the reliable fabrication of probes with nanometer dimensions and the control of tip surface separation. Whilst constant height scanning may suffice for low-resolution (micron scale) measurements of flat samples, the sensitivity of the probe electrode to the tip-surface separation dictates that high-resolution (nanoscale) imaging requires this separation to remain small and constant throughout the measurement. This allows surface activity to be measured independently of topography.

A variety of techniques have been demonstrated that allow a nano-sized electrochemical probe to respond in real time to follow the surface topography. An elegant approach is to combine SECM with atomic force microscopy (AFM).[10] By integrating an addressable electrode onto the tip of an AFM cantilever, the tip-surface separation can be maintained at a fixed value whilst scanning, using the conventional feedback of AFM. Furthermore, this gives the added advantage of simultaneously revealing topographic information about the surface, which can be correlated with the activity map yielded from the electrochemical measurement. A range of SECM-AFM probe fabrication strategies has been demonstrated.[11-17] One particularly successful approach, proposed by Macpherson and co-workers, was to fabricate AFM probes with metal nanowires attached to the tip apex, by using a carbon nanotube as a template.[12] Subsequent insulation and focussed ion beam (FIB) milling of these nanowire probes yielded high aspect ratio tips with an addressable nano-disk electrode at the end. For this type of probe the tip-substrate separation can be controlled for the electrochemical measurements using "lift-mode" in which the tip first traces the topography and subsequently re-scans each line at a fixed distance above the surface.

A drawback of the many fabrication procedures described in the literature is their complexity, which has limited their use to specialist research groups and prevented widespread commercial availability. In this work we explore two alternative approaches to SECM-AFM probe fabrication and highlight the relative merits and limitations.

The first of these approaches involves the use of commercially available high aspect ratio metallic needle probes.[18,19] Ag$_2$Ga alloy needles can be produced with diameters as small as 50 nm by retracting a silver-coated AFM probe from a liquid gallium drop. These probes are of notable use in electric force microscopy (EFM), owing to their electrically conducting nature and the ability to make an electrical connection through the silver coated cantilevers. The probes can be purchased with an insulating Parylene C coating, deposited via chemical vapour deposition. The use of focused ion beam (FIB) milling allows the insulated needles to be cut, exposing the central needle, which can be used as an electrode (see Figure 1a). By adopting the lift-mode approach of Macpherson and coworkers, we report the first application of such probes to combined SECM-AFM.

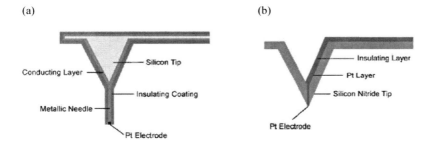

Fig. 1. Schematic depiction of two alternative SECM-AFM probe approaches, shown in cross section. (a) Needle probe. (b) FIB machined probe.

The second fabrication technique investigated takes a rather different approach. One of the most challenging aspects of probe design is the problem of insulation. The electroactive region of the tip must be well defined and the remainder of the probe sufficiently insulated so as to avoid Faradaic processes elsewhere. Coating materials that have been used include Parylene C, silicon nitride, electrophoretic paint and poly(oxyphenylene). The problem is that the coating thickness required is generally of the order of hundreds of nanometres which, except in the more elaborate designs,[11] can compromise the topographical resolution. Here we explore an alternative in which we take a commercial AFM probe and mill *inwards* using

FIB machining. Doing so allows the integration of a working electrode that can be addressed via electrical connection through the interior of the pyramidal tip (see Figure 1b). This way the exposed metal region is a well-defined point at the apex of the tip with nanoscale dimensions, and the remainder of the imaging side of the probe is insulated by the silicon nitride base. A similar approach was adopted by Menozzi and co-workers for the fabrication of EFM probes.[20] An additional requirement for the SECM-AFM application is the necessity to insulate the electrical connection pathway on the rear side of the cantilever. Since the cantilever itself does not affect the imaging resolution, the thickness of insulation added to this part of the probe is not an issue, and so localised insulator deposition can be used. The advantage of this approach as compared to the needle method is, in theory, the improvement in topographical imaging resolution without compromising the electrochemical resolution.

2 Experimental

2.1 Materials

All chemicals and solvents used were of the highest commercially available purity and were used as-received. Hexaamine ruthenium (III) chloride was purchased from Acros Organics and potassium tetrachloroplatinate and chloroplatinic acid from Aldrich. Silver conducting paint was purchased from RS Components. Cathodic electrophoretic paint (Clearclad HSR) was kindly donated by LVH Coatings, Birmingham, UK. Water with a resistivity of not less than 15 MΩcm was taken from an Elga Purelab water purification system. Parylene encapsulated needle probes were purchased from NaugaNeedles LLC, Lousville, KY, USA, with the following specification: 225 μm cantilever length, 2 N/m spring constant, 75 kHz resonant frequency, 5 $\pm$ 3 μm needle length, 100 $\pm$ 50 nm needle diameter, 150 $\pm$ 50 nm thick parylene coating. A range of commercial AFM probe models were tested for suitability in the development of the FIB machining approach. Veeco OTR (Veeco Instruments Europe, France) and Nanoworld PNP-TR (Windsor Scientific, UK) were considered the most successful. Silver and platinum wire were obtained from Goodfellow, Cambridge, UK. A sample grid consisting of 2 μm lines of gold on a silicon substrate was fabricated using e-beam lithography. After masking the required pattern, an adhesion layer of 3 – 5 nm thick titanium was deposited, followed by 50 nm gold.

2.2 Instrumentation

A CH Instruments (Austin, Tx, USA) model 760C bipotentiostat equipped with a CHI200 picoamp booster was used to undertake electrochemical experiments. The picoamp booster was attached to a custom built Faraday cage that was fitted over the AFM scanner and head. AFM measurements were made using a Veeco (Santa Barbara, CA, USA) MultiMode AFM with a Nanoscope IIIa controller. SECM-AFM probes were mounted into a Veeco fluid cell, with electrical connection achieved via the spring loaded metal clasp with the aid of conducting silver paint. Bulk insulation of the probe and clasp was achieved using a combination of nail varnish and superglue (Loctite 407, Farnell Electronic Components Ltd), which was cured at room temperature for several hours. A silver wire quasi-reference/counter electrode was inserted into one of the fluid inputs to the cell. Electron microscopy was undertaken using a Zeiss Supra 40 SEM and FIB machining was achieved using an FEI Nova Nanolab 600 dual beam FIB.

2.3 Probe Development

2.3.1 Needle Probes

A simple three-step procedure was used to modify the commercially available coated needle probes for use in SECM-AFM (Figure 2). Firstly the insulation provided by the parylene coating needed to be improved, since this coating alone was not sufficiently robust for extended use. An improved insulating layer was generated by using a cathodic electrophoretic paint (Clearclad HSR) in addition to the parylene layer. The parylene coated needle probes were immersed in an aqueous solution of the electrophoretic paint (water:paint, 3:1 v/v) and a voltage of -20 V was applied for 4 min, using a Pt wire counter electrode. The probes were rinsed with water and cured at 170 °C for 30 min. After repeating this procedure for a second time, the insulation was tested by immersing the probes into a 1 mM solution of $Ru(NH_3)_6^{3+}$ and undertaking cyclic voltammetry (CV). Whereas typical currents for as-received probes were in the order of tens of pA, after this coating enhancement, only sub-pA currents were observed, indicating a significant improvement in the electrical insulation. Electron microscopy of the probes suggested that the electrophoretic insulation step serves to effectively fill any imperfections in the parylene layer, but does not noticeably increase the needle dimensions.

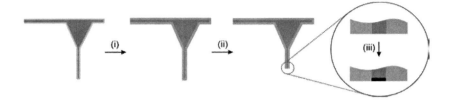

Fig. 2. Schematic of three-step needle probe modification procedure. (i) Coating enhancement, (ii) FIB cutting and (iii) platinum exchange reaction.

The second step in the probe modification to cut the ends off the coated probes using FIB machining, in order to expose the metal core. Using a minimum number of scans to align the probes, a few hundred nm of material was quickly removed from the end of the needles with a beam current of 50 pA.

The final step was to replace the exposed Ag$_2$Ga alloy disk with a more ideal electrode material, namely platinum. This could be achieved by using an electroless galvanic exchange reaction, in which the tip of the needles were immersed in a 10 mM aqueous solution of K$_2$PtCl$_4$ for 2 min:

$$Ag_2Ga + 2PtCl_4^{2-} \rightarrow 2Ag^+ + Ga^{2+} + 2Pt^0 + 8Cl^-$$

A typical example of a finished probe is shown in Figure 3. The pyramidal AFM tip can be seen with a coated needle attached, and a clean cross section cut through the end of the needle. The parylene coating and internal metallic core clearly contrasted, and in this case the total needle radius is close to 360 nm, and the exposed metallic disk has a radius of approximately 160 nm.

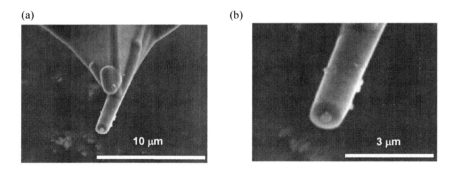

Fig. 3. SEM images of needle probe after modification procedure.

2.3.2 FIB Machined Probes

In this approach, FIB machining was used to mill an aperture through the tip of a silicon nitride AFM probe. FIB-assisted metal deposition was then used to fill the aperture with platinum, yielding an addressable nano-electrode at the tip apex. Silicon nitride probes were used due to the intrinsically electrically insulating nature of this material, and the Veeco OTR and Nanoworld PNP-TR probes in particular had the added advantage of having a hollow geometry, in which the reverse side of the pyramidal tip was recessed (see Figure 4a), which made the FIB milling process considerably less laborious.

The FIB milling procedure involved mounting the cantilevers tip down on a simple silicon substrate, using a thin strip cut from a carbon adhesive disk (Agar Scientific, UK) to raise the tip clear from the substrate. Then, imaging the back of the cantilever, the tip was located and a 30 pA ion beam was used to mill an aperture through the tip. Typically this was achieved in under a minute. Alignment was simple as the edges of the pyramidal tip provided a set of alignment marks running to the apex of the tip. Figure 4b depicts the topside of such a probe with a *ca.* 150 nm diameter aperture successfully milled through the apex. The aperture diameter depends on various parameters, including the beam current, but apertures as narrow as 100 nm were possible with the lowest currents.

Platinum deposition was required both to fill the aperture in order to yield a nano-electrode at the tip apex and in order to provide an isolable conducting path to the point of connection on the probe body. Attempts to fill the aperture using electrochemical deposition of platinum were met with little success, but an alternative method was established based on the work of Menozzi and co-workers, who used beam-assisted platinum deposition to fabricate EFM probes. Essentially this method is ion-beam assisted chemical vapour deposition. An organometallic precursor (methylcyclopentadienyl-tri-methyl-platinum) is used to provide the platinum material and is introduced into the FIB chamber as a gas via a needle positioned 50 μm above the sample surface. This gas is then dissociated by the ion beam, leaving behind a platinum rich deposition that also contains some gallium from the ion beam, and carbon from the precursor. Using this technique, it was possible to fill the aperture with platinum initially from the rear side of the cantilever and completing the deposition from the tip side. The result can be seen in Figures 4c and 4d, which depict the filled tip from both sides of the cantilever. Connection lines were also deposited along the cantilever as depicted in Figure 4e. These connections were extended up the edge of the probe and onto the top of the main body, terminating with a large (*ca.* 200 square micron) connection pad (see Figure 4f). Note that some commercial probes have multiple cantilevers attached as seen in Figure 4f, but the unmodified cantilevers were typically removed at the end of the probe modification process.

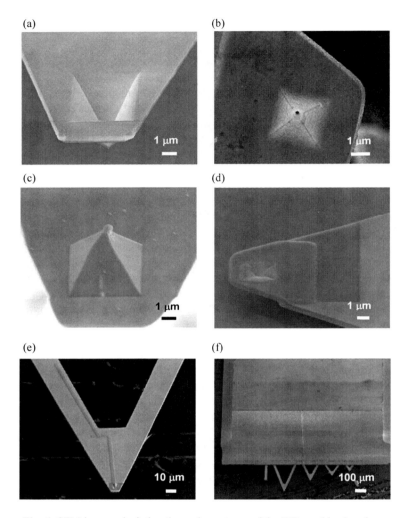

Fig. 4. SEM images depicting the various stages of the FIB machined probe approach. (a) Hollow recess on reverse side of cantilever, (b) aperture milled through tip, (c) deposited Pt protruding from pyramidal tip, (d) Pt filled recess on reverse side of cantilever, (e) deposited Pt connection lines on reverse side of cantilever and (f) Pt connection line and contact pad on probe body.

Insulation of the exposed platinum connection lines was necessary in order to ensure that all electrochemical currents originate only from the nanoelectrode positioned at the apex of the pyramidal tip. In a similar fashion to above, the connection lines and exposed platinum on the rear side of the cantilever was covered using a beam-assisted procedure to locally deposit a layer of silicon oxide. For this to be successful, it was first necessary to isolate these connection lines from the gold layer that covers the cantilever and probe body (this layer is typically added by the manufacturer to improve reflectivity). This was achieved by using the FIB to mill away the gold from the surface at all regions where it was connected to the deposited platinum in order to isolate the conducting pathway.

The insulating layer was deposited using the same method as the platinum. In this case the precursor consisted of two-parts, TEOS (tetraethylorthosilicate) and water vapour. When the TEOS chemical is dissociated by the ion beam two products are formed, ethanol and silicon oxide. The ethanol was pumped away by the vacuum system and silicon oxide deposited. The deposit does contain some gallium from the beam and so is not as insulating as pure silica, but was deemed sufficient. The thin lines of platinum were covered with wider strips of insulator, and using the SEM it was possible to see how complete the coverage was. The deposited material is inherently low contrast in the SEM, whereas the platinum is very high contrast.

3 Results and Discussion

3.1 Electrochemical Characterization

Preliminary assessment of the electrochemical behavior of both types of SECM-AFM probe was carried out using cyclic voltammetry. The probes were sealed into an AFM fluid cell and immersed into a 5 mM aqueous solution of $Ru(NH_3)_6Cl_3$ containing 0.1 M KNO_3. We begin with the behaviour of the needle probes before moving onto the FIB machined probes.

A typical CV of a fully modified needle probe is depicted in Figure 5a, which shows a current plateau emerging at -0.4 V *vs* Ag, indicative of the Faradic reduction of the Ru^{3+} complex to Ru^{2+}. Whilst a steady-state diffusion limited current is difficult to assign precisely, due to the sloping nature of the CV at more negative potentials, we estimate this is in the region of 230 pA. If one assumes a disk geometry, the electrochemical radius, r, of that disk can be estimated from the limiting current, i_{lim}:[21]

$$i_{lim} = 4FrDc \tag{1}$$

where F is Faraday's constant, D is the diffusion coefficient of the Ru^{3+} complex (taken as 8.9×10^{-6} cm^2 s^{-1})[22] and c is its concentration. From this we estimate an electrochemical radius close to 130 nm. Control experiments in 0.1 M KNO_3 alone suggested negligible background currents. It is noteworthy that the electrochemical radius determined is slightly smaller than the radius of the metal core shown in Figure 3. We can attribute this to either the blocking of the electrode surface by material removed during the FIB procedure, or the incomplete Pt exchange of the exposed core metal, leading to a partially inactive electrode.

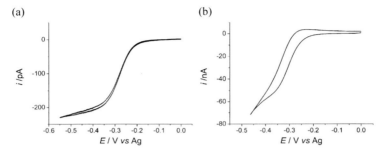

Fig. 5. Cyclic voltammetry of SECM-AFM probes at 20 mV/s. (a) Modified needle probe submerged in 5 mM Ru(NH₃)₆Cl₃/0.1 M KNO₃. (b) FIB machined probe submerged in 1 mM Ru(NH₃)₆Cl₃/0.1 M KNO₃.

In order to verify that the sole source of the Faradaic current was the exposed disk electrode, silver electrodeposition was carried out by immersing a probe in a Ag-NO₃ solution and applying a bias of -0.55 V. Subsequent inspection of the probe using SEM indicated significant silver growth at the tip of the needle, but no evidence of deposition elsewhere. This further confirmed that the coating provided sufficient electrical insulation, such that electrochemical control is confined to purely the tip of needle.

The same voltammetric procedure as above was carried out on the FIB machined probes and a typical CV is depicted in Figure 5b wherein an inflexion in the current occurs at a potential of approximately –0.35 V *vs.* Ag, which can again be attributed to the reduction of Ru³⁺ to Ru²⁺. On the reverse scan there is a discernable peak due the re-oxidation of Ru²⁺ indicating that there is a transient element to the CV at this scan rate (as opposed to a pure steady state CV in which a flat plateau is expected). Most importantly, the Faradaic current measured is close to 60 nA. If we assume a disk geometry, as above, substitution into equation (1) yields a disk radius of *ca.* 170 μm. This is a clear indication that there is a current leakage problem, consistent with either insufficient/incomplete TEOS insulation, or poor isolation between the deposited Pt connection lines and the remaining gold layer on the surface of the cantilever. Given that a disk with a radius of the order of one hundred microns could easily cover the whole cantilever, it is most likely that poor isolation is the issue and that a connection still exists between the deposited platinum and the gold coating on the cantilever. This was confirmed through silver electrodeposition experiments. Further investigations are underway to find the source of the isolation problem but are beyond the scope of the current article. So at this stage it was concluded that the FIB machining approach, whilst innovative, is not capable at present of producing a viable solution to the SECM-AFM probe fabrication issue. As such, for the final phase of testing, we consider only the needle probes.

3.2 SECM-AFM Imaging

The suitability of the needle probes for SECM-AFM imaging was demonstrated with the aid of the test substrate consisting of well-defined regions of deposited gold. Measurements were undertaken in a solution of 5 mM Ru(NH$_3$)$_6$Cl$_3$/0.1 M KNO$_3$ with a constant bias of -0.45 V *vs* Ag applied to the tip. Lift mode was employed, wherein two scans were performed for each line, the first to determine the topography, and the second to scan above that topography at a fixed distance of 200 nm. The test substrate was biased at 0 V *vs.* Ag such that the Ru^{2+} generated by the probe was quickly reoxidized to Ru^{3+}, leading to an enhanced current response (positive feedback) in the regions of the gold. SECM-AFM images recorded are shown in Figure 6, which indicates both the topographical (a) and electrochemical (b) data. The gold features are clearly resolved in the topographical imaging, and there is good contrast in the electrochemical response between the conducting and non-conducting regions of the sample. The line scans shown indicate that the height of the gold features are in the region of 50 nm (as expected from the substrate fabrication procedure) and that current differentiation between the different regions is in the order of nA. This is slightly larger than the currents measured in the preliminary voltammetry which we attribute to the slow degradation of the probe insulation over extended periods of use. Improvements to probe durability are a focus of ongoing work.

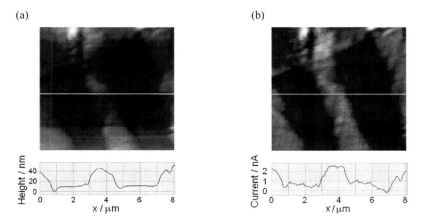

Fig. 6. SECM-AFM imaging of gold patterned substrate using needle probe immersed in 5 mM Ru(NH$_3$)$_6$$^{3+}$/0.5 M KNO$_3$. (a) Topographical (height) image and line scan. (b) Electrochemical image and line scan. Scan rate = 0.2 Hz, probe potential −0.45 V *vs.* Ag.

4 Conclusions and Outlook

In this work we have explored two different novel approaches to SECM-AFM probe fabrication, which have met different degrees of success in testing. Whilst, in theory, both approaches have the potential to allow improvement in the resolution of the SECM-AFM technique, the needle probe method was found to yield the most promising results. In both cases continued development should permit further improvements to resolution and durability. For the needle probes this can be achieved through decreasing both the needle diameter and the thickness of coating required for electrical insulation around the needle. In the case of the FIB machining approach, further work is required to tackle the obstacle of poor electrical isolation, but the electrochemical resolution is only limited by the minimum aperture dimensions, which is close to 100 nm.

The development of SECM-AFM probes continues to be a very challenging problem, and as such the use of this technique is not widespread. However, with improvements to resolution, ease of fabrication and the eventual commercial availability of such probes, the full potential of this powerful approach may be realized, and will ultimately serve to improve our understanding of interfacial processes at the nanoscale.

Acknowledgments

This work was undertaken as part of the Innovation R&D Programme of the Department for Innovation, Universities and Skills (DIUS). The authors wish to thank Doctor Patrick Nicholson of NPL for his contribution towards the planning and development of this project. We are also grateful to Mehdi Yazdanpanah of NaugaNeedles LLC for assisting with the needle probe development.

References

1. Amemiya S, Bard AJ, Fan FRF, Mirkin MV, Unwin PR (2008) Scanning electrochemical microscopy. Annu Rev Anal Chem 1:95-131.
2. Bard AJ, Mirkin MV (eds) (2001) Scanning electrochemical microscopy. Marcel Dekker, New York
3. Eckhard K, Etienne M, Schulte A, Schuhmann W (2007) Constant-distance mode ac-secm for the visualisation of corrosion pits. Electrochemistry Communications 9 (7):1793-1797.
4. Macpherson JV, Unwin PR (1996) Scanning electrochemical microscope-induced dissolution: Theory and experiment for silver chloride dissolution kinetics in aqueous solution without supporting electrolyte. Journal of Physical Chemistry 100 (50):19475-19483
5. Wain AJ, Zhou FM (2008) Scanning electrochemical microscopy imaging of DNA microarrays using methylene blue as a redox-active intercalator. Langmuir 24 (9):5155-5160.
6. Bard AJ, Li X, Zhan W (2006) Chemically imaging living cells by scanning electrochemical microscopy. Biosensors & Bioelectronics 22 (4):461-472.
7. Nicholson PG, Zhou S, Hinds G, Wain AJ, Turnbull A (2009) Electrocatalytic activity mapping of model fuel cell catalyst films using scanning electrochemical microscopy. Electrochimica Acta 54 (19):4525-4533.

8. Katemann BB, Schuhmann T (2002) Fabrication and characterization of needle-type pt-disk nanoelectrodes. Electroanalysis 14 (1):22-28
9. Laforge FO, Velmurugan J, Wang YX, Mirkin MV (2009) Nanoscale imaging of surface topography and reactivity with the scanning electrochemical microscope. Analytical Chemistry 81 (8):3143-3150.
10. Gardner CE, Macpherson JV (2002) Atomic force microscopy probes go electrochemical. Analytical Chemistry 74 (21):576A-584A
11. Kueng A, Kranz C, Mizaikoff B, Lugstein A, Bertagnolli E (2003) Combined scanning electrochemical atomic force microscopy for tapping mode imaging. Applied Physics Letters 82 (10):1592-1594.
12. Burt DP, Wilson NR, Weaver JMR, Dobson PS, Macpherson JV (2005) Nanowire probes for high resolution combined scanning electrochemical microscopy - atomic force microscopy. Nano Letters 5 (4):639-643.
13. Macpherson JV, Unwin PR (2000) Combined scanning electrochemical-atomic force microscopy. Analytical Chemistry 72 (2):276-285
14. Abbou J, Demaille C, Druet M, Moiroux J (2002) Fabrication of submicrometer-sized gold electrodes of controlled geometry for scanning electrochemical-atomic force microscopy. Analytical Chemistry 74 (24):6355-6363.
15. Anne A, Cambril E, Chovin A, Demaille C, Goyer C (2009) Electrochemical atomic force microscopy using a tip-attached redox mediator for topographic and functional imaging of nanosystems. Acs Nano 3 (10):2927-2940.
16. Fasching RJ, Tao Y, Prinz FB (2005) Cantilever tip probe arrays for simultaneous secm and afm analysis. Sensors and Actuators B-Chemical 108 (1-2):964-972.
17. Pust SE, Salomo M, Oesterschulze E, Wittstock G (2010) Influence of electrode size and geometry on electrochemical experiments with combined secm-sfm probes. Nanotechnology 21 (10).
18. Yazdanpanah MM, Harfenist SA, Safir A, Cohn RW (2005) Selective self-assembly at room temperature of individual freestanding ag2ga alloy nanoneedles. Journal of Applied Physics 98 (7).
19. http://www.nauganeedles.com/.
20. Menozzi C, Gazzadi GC, Alessandrini A, Facci P (2005) Focused ion beam-nanomachined probes for improved electric force microscopy. Ultramicroscopy 104 (3-4):220-225.
21. Bard AJ, Faulkner LR (2001) Electrochemical methods: Fundamentals and applications. 2nd edn. Wiley, New York
22. Birkin PR, SilvaMartinez S (1997) Determination of heterogeneous electron transfer kinetics in the presence of ultrasound at microelectrodes employing sampled voltammetry. Analytical Chemistry 69 (11):2055-2062

Electrochemical Synthesis of Nanostructured Pd-based Catalyst and Its Application to On-Chip Fuel Cells

Satoshi Tominaka

World Premier International Research Center for Materials Nanoarchitectonics (MANA), National Institute for Materials Science (NIMS), Namiki 1-1, Tsukuba, Ibaraki 305-0044, Japan. E-mail: TOMINAKA.Satoshi@nims.go.jp

Abstract

Pd-based catalyst is a promising alternative to Pt-based catalysts in terms of their comparable activity for the oxygen reduction reaction and significantly high fuel tolerance. Here, the mesoporous PdCu catalyst was synthesized directly on current collectors by the combination of electrodeposition and dealloying. This synthetic method is suitable for micro-fuel cells. The PdCu catalyst thus obtained was a 400-nm thick mesoporous film composed of 10–40-nm pores and 10–20-nm thick ligaments. The composition was determined to be $Pd_{88}Cu_{12}$ by X-ray spectroscopy. The surface was determined to be a polycrystalline Pd surface by cyclic voltammetry. Interestingly, the current peaks associated with hydrogen adsorption–desorption reactions were significantly sharp, suggesting large Pd(111) terrace sites were formed. TEM observation showed that the ligaments of the mesoporous structure were not composed of tiny crystallites but of relatively large string-like crystallites, supporting the consideration described above. Since similar feature was also observed in the case of PdCo catalyst synthesized by a similar method, this unique surface feature is probably typical for dealloyed materials.

Keywords: Electrocatalyst, PdCu catalyst, Mesoporous catalyst, Dealloying, Electrodeposition

1. Introduction

There is an increasing demand for the development of effective electrodes to improve the performance of electrochemical devices, including fuel cells, batteries, sensors and electrolysis apparatuses, which are attractive to create a sustainable society. Materials as well as their synthetic process should be chosen in accordance with the intended use. For example, direct alcohol fuel cells (DAFCs), which operate on alcohol and oxygen, need electrocatalysts, generally, composed of Pt or its alloys. Though the Pt-based catalysts exhibit best activity for the oxygen reduction reaction (ORR), Pt has a severe problem that it can catalyze not only ORR but also the alcohol oxidation reaction. Thus, in order to avoid the alcohol oxidation on ORR catalyst, which decreases power as well as fuel efficiency, two electrodes of DAFCs must be chemically separated by an ion-conductive membrane. This drawback hampers the miniaturization of DAFCs and complicates their system.

Since Pd-based catalysts (*e.g.*, PdCo and PdCu) are attractive as electrocatalysts for DAFCs in terms of their high ORR activity comparable to Pt-based ones and their high tolerance to alcohol [1-7], on-chip fuel cells of a membraneless design have been developed by this author and coworkers as a promising micro-power source for miniaturized devices [8-12]. In order to increase electric current from the device, porous electrodes having a large effective surface area are needed. Roughly speaking, the surface area of a porous electrode can reach a value comparable to those of nanoparticles when the pore size gets to a mesoscale (*e.g.*, 20–50 m^2 g^{-1} for mesoporous Pt). For the selective deposition onto microelectrodes (submillimeter range) of the on-chip fuel cells, electrodeposition has been used, because this technique can synthesize metals and alloys selectively onto current collectors, making them inherently electron-conductive.

Furthermore, for the synthesis of mesoporous catalyst layers directly onto tiny current collectors of on-chip fuel cells, this author and coworkers used electrodeposition in combination with dealloying. Dealloying refers to the selective dissolution of less-noble element(s) out of an alloy and forms unique sponge-like porous structure of the noble components with keeping their macrostructures and without template [13-15]. The unique porous structure is formed by a competition of two processes: dissolution of less-noble component (*i.e.*, pore formation) and surface diffusion of nobler component to aggregate into two-dimensional clusters (*i.e.*, surface passivation) [13,14]. For example, a coral-reef-like nanostructure of PdCo alloy was synthesized by the electrodeposition and dealloying [16], and such a mesoporous PdCo catalyst was found to exhibit a higher ORR activity than Pt in the potential range for fuel cell operation [17].

The origin of the superior activity has not been clarified yet. In order to investigate if the superiority originated from dealloying process, *e.g.*, some unique property of the surface created by the synthesis, similar but still different system of PdCu is tested. Cu has an atomic radius of 1.28 Å, which is similar to that of Co

(1.25 Å) but is smaller than that of Pd (1.37 Å), causing similar lattice contraction effect, which is considered to dominate the positive alloying effect of Pd-based ORR catalysts [18]. From the viewpoint of electron orbitals, Cu is richer in d-orbital electrons ([Ar] $3d^{10} 4s^1$) than Co ([Ar] $3d^7 4s^2$), this probably causing different electronic effects when alloyed with Pd ([Kr] $4d^{10}$). In view of synthesis, the previous PdCo electrodeposition needed a quite complicated solution containing multiple ligands, but PdCu is expected to deposit from a simpler solution (details are discussed in Section 3.1) [19]. Thus, here the synthesis of mesoporous PdCu films is reported for the detailed discussion on the dealloyed Pd-based electrocatalyst. The PdCu is synthesized by the combination of electrodeposition and dealloying as illustrated in Fig. 1.

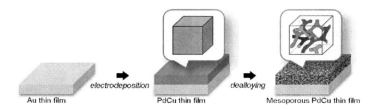

Fig. 1 Schematic illustration for representing the synthesis of mesoporous PdCu electrode by the combination of electrodeposition and dealloying.

2. Experimental

PdCu alloy was first electrodeposited at 0.1 V in a solution containing 10 mM K_2PdCl_4, 10 mM $CuSO_4$ and 0.5 M H_2SO_4 for 10 min. Immediately after the deposition, the deposit was dealloyed at 0.55 V for 10 min in a 0.5 M H_2SO_4 solution deaerated with pure Ar. All the electrochemical experiments were conducted using an electrochemical instrument (HZ-5000 from Hokuto Denko) at room temperature. The cell was a conventional three-electrode cell. The working electrode was a 200-nm Au layer deposited by electron-beam evaporation (1.5 Å cm^{-1}) on a silicon wafer covered with a 200-nm SiO_2 layer, on which a 20-nm Ti layer (1.0 Å cm^{-1}) was deposited to improve adhesion between Au and SiO_2. The working electrode area was defined as 5×5 mm^2 with Capton tape. The reference and the counter electrodes were a silver/silver chloride (Ag/AgCl) electrode and a Pt wire, respectively. All the chemicals were purchased from Wako Pure Chemical Industries, Ltd.

The microstructure was observed with a field-emission scanning electron microscope (FE-SEM; S-4800 from Hitachi) at 15 kV and also with a transmission electron microscope (TEM; JEM-2100F from JEOL) at 200 kV. The high resolution TEM images were analyzed using a software, ImageJ, in order to determine lattice distance and to obtain fast Fourier transform (FFT) patterns. The FFT patterns were analyzed using a simulation software, ReciPro. The bulk composition was determined by energy dispersive X-ray (EDX) spectroscopy using a detector (EMAX from Horiba) attached to the SEM. The electrochemical response was analyzed by cyclic voltammetry scanned in a 0.5 M H_2SO_4 deaerated with pure Ar.

3. Results and discussion

3.1 Electrodeposition of PdCu alloys and the following dealloying process

For the better understanding of the synthesis, the underpotential co-deposition of PdCu alloy was confirmed by cyclic voltammetry as follows. In a solution containing Cu^{2+} ions and/or $PdCl_4^{2-}$ ions, stable CVs were obtained as shown in Fig. 2. The CV in the PdCu solution was obviously different from those obtained in the solution containing one of the salts and also from the sum of them, indicating a strong interaction between Pd atoms and Cu atoms during the deposition. Such a strong interaction is known as phenomena, underpotential deposition, in the field of electrochemistry. Cu is known to deposit on Au, Pt and Pd even in the potential range of more positive than Nernst potential for Cu deposition, *i.e.*, in the potential range where bulk Cu is not deposited.

Actually, in the case of the Cu^{2+} solution used for the test shown in Fig. 2a, since the Nernst potential is calculated as 0.084 V *vs.* Ag/AgCl, the reduction peak located around 0.1 V and the oxidation peaks at 0.15 V and 0.27 V were attributable to underpotential deposition of Cu on the Au electrode. Since it was reported that Cu underpotential deposition on Pd commenced around 0.4 V *vs.* Ag/AgCl, the reduction current increase observed for the PdCu solution compared with the solution containing only $PdCl_4^{2-}$ ions is reasonably attributable to the underpotential co-deposition of Cu with Pd. Likewise, the oxidation current observed below 0.4 V at the positive scan indicates that some portion of co-deposited Cu was selectively dissolved.

Based on these considerations on the CVs, potentials for synthesizing nanostructured PdCu catalyst, *i.e.*, the deposition potential and the dealloying potential, were determined to be 0.1 V and 0.55 V, respectively. Fig. 3 shows the current profiles of the deposition process and the dealloying process. The total charge for the deposition was *ca.* 0.96 mC cm^{-2}, and that for the dealloying was *ca.* 0.30 mC cm^{-2}. Assuming that their Coulombic efficiencies were as high as 1.0 and that Cu was dissolved as Cu^{2+}, there values are consistent with dissolution amount of *ca.* 31 at% of the whole deposit.

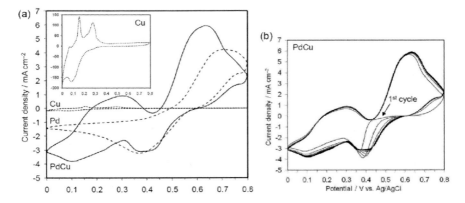

Fig. 2 Cyclic voltammograms scanned at 50 mVs^{-1} in a solution containing (solid line) 10 mM K$_2$PdCl$_4$, 10 mM CuSO$_4$ and 0.5 M H$_2$SO$_4$, (dashed line) 10 mM K$_2$PdCl$_4$ and 0.5 M H$_2$SO$_4$, or (dotted line) 10 mM CuSO$_4$ and 0.5 M H$_2$SO$_4$. (a) 10th cycles are shown. (b) The first 10 cycles for the PdCu deposition.

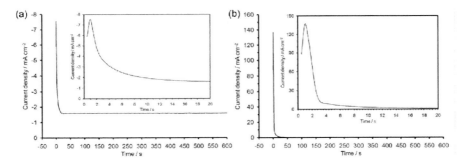

Fig. 3 Current profiles of (a) the electrodeposition process of the PdCu alloy at 0.1 V vs. Ag/AgCl in a solution containing 10 mM K$_2$PdCl$_4$, 10 mM CuSO$_4$ and 0.5 M H$_2$SO$_4$, and (b) the dealloying process at 0.55 V in a 0.5 M H$_2$SO$_4$ deaerated with pure Ar.

3.2 Microstructure of the PdCu layer

The PdCu layer thus obtained was a mirror-like shiny silver film. This optical property probably resulted from the flat outer surface of the film (Figs. 4a and 4b). At higher magnification, the SEM images clearly show that the PdCu layer was mesoporous. The pores were in the range of 10–40 nm, and the ligaments were 10–20 nm thick. Since the as-deposited PdCu was indeed a dense non-porous film (data not shown), the mesopores are judged to be formed by the dealloying process.

The compositions of the as-deposited PdCu film and the dealloyed one were respectively determined to be Pd$_{53}$Cu$_{47}$ and Pd$_{88}$Cu$_{12}$ by the EDX measurements. The composition change corresponds to the dissolution amount of *ca.* 85 at% of deposited Cu or *ca.* 40% of the whole deposit. Compared with the value calculated from deposition–dealloying charges (*i.e.*, 31 at%), this value, *ca.* 40%, is larger and probably suggests that some of Cu atoms were dissolved as Cu$^+$.

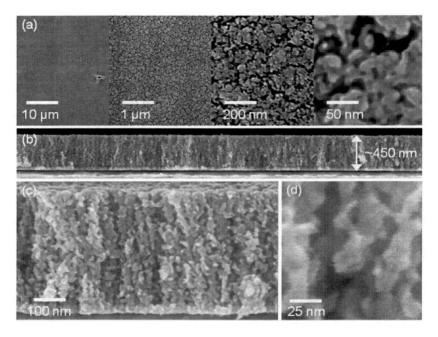

Fig. 4 SEM images of the synthesized mesoporous PdCu film: (a) plan views and (b-d) cross-sectional views at different magnifications.

The TEM observation confirms that the ligaments were 10–20-nm thick (Fig. 5). In the high resolution images, lattice fringes whose lattice spacing of *ca.* 0.22 nm were clearly observed. This spacing value was also confirmed by the fast Fourier transform (FFT) patterns, which also show other spots for different frequencies: one group corresponds to the lattice spacing of *ca.* 0.19 nm, and the other group corresponds to the lattice spacing of *ca.* 0.14 nm. These values are quite consistent with those of pure fcc-Pd lattice (lattice constant 3.908 Å; PDF no. 01-087-0643) observed from <110> direction, where lattice spacing values are 0.226 nm for {111}, 0.138 nm for {220}, and 0.195 nm for {200}. Though those distance values maybe indicate a slight lattice contraction of Pd lattice due to alloying with Cu (lattice constant 3.610 Å for fcc-Cu), detailed analysis is still needed for concluding this point.

The ligament exhibits the same FFT patterns along with the length direction from B point to C point, this meaning that the ligament was not composed of small crystallites but of a string-like crystallite. This fact is consistent with the result obtained for the mesoporous PdCo similarly synthesized by the combination of electrodeposition and dealloying [17]. Thus, such an extensive crystallite growth is considered to be a unique aspect of dealloyed materials. Since the relatively-few crystal grain boundaries may result in a unique surface having large terrace sites, such dealloyed materials are of great interest in terms of catalysts.

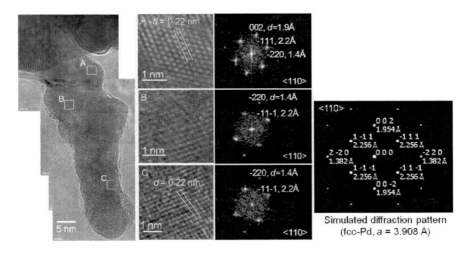

Fig. 5 TEM images of a ligament of the synthesized mesoporous PdCu film. (A-C) High resolution images of the selected area and their corresponding FFT patterns. The right diffraction pattern was obtained by simulation.

3.4 Electrochemical property of the mesoporous PdCu electrode

Next, the surface of the mesoporous PdCu electrode was analyzed by cyclic voltammetry (CV) as shown in Fig. 6. It is known that CV trace of a nanostructured Pd electrode shows hydrogen adsorption–desorption peaks below 0.1 V vs. Ag/AgCl in addition to the large hydrogen absorption–desorption currents usually observed for Pd electrodes around −0.2 V [20]. This feature holds true for the CV of the mesoporous PdCu electrode, which exhibited sharp redox peaks and broad less-reversible peaks for the hydrogen adsorption–desorption reactions (Fig. 6b) [21]. The sharpest peaks located at 0.03 V were assigned to the reactions on {111} faces, and the broad peaks located around −0.05 V to the reaction on {100} faces (Fig. 6b) [17]. The obvious sharpness of the peaks probably indicates that large {111} terrace sites were formed. Since such sharp peaks were not reported in the case of nanoparticles of Pd and Pd alloys as far as this author knows, it is considered that they originate from a unique feature of mesoporous Pd-based materials synthesized by dealloying.

These exposed facets were also confirmed by the analysis of oxide formation current peaks in the range of 0.5–1.0 V (see Figs. 6a and 6b). Itaya et al. reported that the CVs of Pd single crystal electrodes scanned in 0.05 M H_2SO_4 at 10 mV s^{-1} exhibited current peaks associated with surface oxide formation (Pd-OH or PdO) located at 0.9 V vs. Ag/AgCl for Pd(111), 0.70 V for Pd(100) and 0.65 V for Pd(110) [22]. Thus, the peak located at 0.86 V (Fig. 6a) was assigned to {111} faces, the peak located at 0.69 V to {100} faces, and the peak located at 0.61 V (Fig. 6b) to {110} faces. The peak located at 0.57 V (Fig. 6b) may be attributed to some high index facets or surface defect sites.

As discussed above, the CV trace of the mesoporous PdCu indicates that the surface was similar to a polycrystalline Pd electrode. In view of this, roughness factor, which is defined as the ratio of the electrochemically active surface area to the geometrical surface area, of the mesoporous PdCu can be evaluated using a method previously reported for such surface [20]. The roughness factor was determined from the double layer capacitance (2.36 mF cm^{-2} at 0.29 V) to be ca. 27. This value indicates that the electrode surface in the mesopores was electrochemically active.

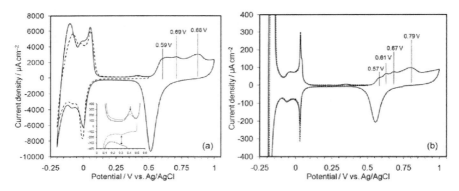

Fig. 6 Cyclic voltammograms of the synthesized mesoporous PdCu film in a 0.5 M H_2SO_4 solution in the potential ranges of –0.2 to 1.0 V (solid lines) and –0.2 to 0.5 V (dotted lines) at (a) 50 mVs^{-1} and (b) 1 mVs^{-1}. The inset in (a) shows enlarged voltammograms.

Interestingly, additional redox peaks located at 0.40 V for the oxidation current and at 0.33 V for the reduction current were observed (see the inset, Fig. 6a). These peaks were stably obtained even after 2 h of the CV measurement, indicating that they were not originating from surface impurities and/or redeposition and the corresponding stripping of residual Cu ions. Indeed, peak potentials for the Cu underpotential deposition were reported to be located at 0.30 V *vs.* Ag/AgCl for (100) faces and 0.27 V for (111) faces, thus the redox peaks obtained were found not to be such. The peaks might result from adsorbed sulfate or oxidation of some defect sites, but further detailed analyses are needed to clarify the origin.

4. Conclusion

Mesoporous PdCu film was successfully synthesized by the underpotential codeposition of Pd and Cu followed by dealloying. This mesoporous $Pd_{88}Cu_{12}$ film had a polycrystalline surface similar to pure Pd. Interestingly, the current peaks associated with hydrogen adsorption–desorption reactions were significantly sharp, suggesting large terrace sites of Pd(111) were formed. TEM observation showed that the ligaments of the mesoporous structure were not composed of tiny crystallites but of relatively large string-like crystallites, supporting the consideration on the formation of large terrace sites. Probably, this feature is typical for dealloyed materials.

In addition to these well-characterized features, small redox peaks which had not been observed for the previous mesoporous PdCo films were observed in the potential range for double layer region in CV. In future, analysis of these peaks and characterization as catalyst will be carried out.

Acknowledgements

This work was in part supported by World Premier International Research Center Initiative (WPI Initiative) on "Materials Nanoarchitronics", MEXT, Japan. The author thanks to Dr. Y. Nemoto (NIMS, Japan) for the experimental support on the TEM works, and to MANA foundry (NIMS, Japan) for the experimental supports on the preparation of the Au-deposited Si wafers.

References

1. Fernandez JL, Raghuveer V, Manthiram A, Bard AJ (2005) J Am Chem Soc 127 (38):13100-13101. doi:Doi 10.1021/Ja0534710
2. Lee K, Savadogo O, Ishihara A, Mitsushima S, Kamiya N, Ota K (2006) J Electrochem Soc 153 (1):A20-A24. doi:Doi 10.1149/1.2128101
3. Mustain WE, Kepler K, Prakash J (2006) Electrochem Commun 8 (3):406-410. doi:DOI 10.1016/j.elecom.2005.12.015

4. Mustain WE, Kepler K, Prakash J (2007) Electrochim Acta 52 (5):2102-2108. doi:DOI 10.1016/j.electacta.2006.08.020
5. Raghuveer V, Manthiram A, Bard AJ (2005) J Phys Chem B 109 (48):22909-22912. doi:Doi 10.1021/Jp054815b
6. Savadogo O, Lee K, Oishi K, Mitsushima S, Kamiya N, Ota KI (2004) Electrochem Commun 6 (2):105-109. doi:DOI 10.1016/j.elecom.2003.10.020
7. Wang XP, Kariuki N, Vaughey JT, Goodpaster J, Kumar R, Myers DJ (2008) J Electrochem Soc 155 (6):B602-B609. doi:Doi 10.1149/1.2902342
8. Yeston J (2009) Science 325:1321
9. Tominaka S, Nishizeko H, Ohta S, Osaka T (2009) Energy Environ Sci 2:849-852. doi:10.1039/b906216e
10. Tominaka S, Nishizeko H, Mizuno J, Osaka T (2009) Energy Environ Sci 2 (10):1074-1077. doi:Doi 10.1039/B915389f
11. Tominaka S, Ohta S, Obata H, Momma T, Osaka T (2008) J Am Chem Soc 130 (32):10456-10457. doi:Doi 10.1021/Ja8024214
12. Osaka T, Tominaka S (2008) Asia Materials (on-line): 28th Oct.
13. Erlebacher J, Aziz MJ, Karma A, Dimitrov N, Sieradzki K (2001) Nature 410 (6827):450-453
14. Rugolo J, Erlebacher J, Sieradzki K (2006). Nat Mater 5 (12):946-949. doi:Doi 10.1038/Nmat1780
15. Wagner K, Brankovic SR, Dimitrov N, Sieradzki K (1997) J Electrochem Soc 144 (10):3545-3555
16. Tominaka S, Nakamura Y, Osaka T (2010) J Power Sources 195 (4):1054-1058. doi:DOI 10.1016/j.jpowsour.2009.08.082
17. Tominaka S, Hayashi T, Nakamura Y, Osaka T (2010) J Mater Chem DOI: 10.1039/c0jm00973c
18. Suo YG, Zhuang L, Lu JT (2007) Angew Chem Int Edit 46 (16):2862-2864. doi:DOI 10.1002/anie.200604332
19. Milhano C, Pletcher D (2008) J Electroanal Chem 614 (1-2):24-30. doi:DOI 10.1016/j.jelechem.2007.11.001
20. Tominaka S, Momma T, Osaka T (2008) Electrochim Acta 53 (14):4679-4686. doi:DOI 10.1016/j.electacta.2008.01.069
21. Baldauf M, Kolb DM (1993) Electrochim Acta 38 (15):2145-2153
22. Sashikata K, Matsui Y, Itaya K, Soriaga MP (1996) J Phys Chem 100 (51):20027-20034

Characterization and Synthesis of PtRu/C Catalysts for Possible use in Fuel Cells

Eleanor Fourie, Gary Pattrick, Elma van der Lingen

Advanced Materials Division – Mintek, Private Bag X3015, Randburg, 2125, South Africa.

Tel.: +27 11 709 4316. E-mail: eleanorf@mintek.co.za.

Abstract Mixed metal nanocatalysts containing platinum and ruthenium on a carbon support were synthesized for possible use in direct methanol fuel cells (DMFC). The synthesis of these catalysts, as described in the literature, is a standard impregnation method of the metal salts, followed by reduction. The reduction can be carried out in a number of ways - either by the addition of a liquid reducing agent or by passing a reducing gas over the filtered and dried catalyst. In this study, PtRu/C catalysts were prepared by reduction with various possible reducing agents, i.e. formaldehyde, formic acid and hydrogen gas at relatively low (923 K) and high (1 173 K) temperatures. The catalyst material was tested by transmission electron microscopy (TEM), X-ray diffraction (XRD) and electrochemical methods in order to determine the particle size, alloy formation and catalytic activity of the material. It was found that milder reducing agents led to smaller particle sizes of the metal particles on the carbon support. Reduction conditions were also found to significantly influence the properties of the catalyst. A variety of different metallic, hydrated and oxide species of the precious metal particles are possible. Temperature programmed reduction (TPR) was used to investigate the relative oxidation state of the metal particles.

1 Introduction

As the need for alternative energy resources increases, fuel cells have become a major focus of research incorporating various fields of science. Many types of fuel cells are being investigated, all with different properties and different applications. Direct methanol fuel cells (DMFC) oxidize methanol as fuel at the anode and reduce oxygen from air at the cathode. The advantages of using methanol as a fuel are its high theoretical energy density, and because it is a liquid at room temperature it is easily handled, transported and stored. DMFCs are operated under moderate temperatures (< 370 K) and are considered ideal for providing portable power for small standby power devices, especially in remote areas. Portable power systems could find application in domestic, commercial, military, telecoms, IT and other situations, and are expected to be the early adopters of fuel cell technology since these markets are more able to tolerate the higher initial implementation costs. [1, 2]

It has been found experimentally that the best anode catalyst for methanol oxidation contains platinum and ruthenium in a 50:50 mole ratio. The full mechanism of this catalyst is still under investigation. Initially it was thought that a bimetallic alloy of ruthenium and platinum would be preferential, but this was shown not to be the case. [3] It has generally been accepted that these catalysts can contain a variety of different metallic, hydrated and oxide species of Pt and Ru. Recent studies seem to agree on a bifunctional mechanism for the overall oxidation of methanol. [3-8] Firstly metallic Pt on the surface is responsible for the oxidation of methanol to carbon monoxide, which is very strongly adsorbed on the Pt surface. This, however, leads to poisoning of the Pt surface, but the CO can be removed through an oxygen-transfer step by Ru-OH electrogenerated from hydrous ruthenium oxide ($RuO_2.xH_2O$). This bimetallic mechanism is summarized by eq 1. [1]

$$Pt(CO)_{ad} + Ru(OH)_{ad} \rightarrow CO_2 + H^+ + e^- \tag{1}$$

In this study, various reduction techniques were used to synthesize a number of catalysts, all with the same precious metal loading and carbon support. A variety of techniques were used to fully investigate the material properties. Electrochemical methods were used to test the catalytic performance. The synthetic procedure was found to significantly influence the catalytic performance, and attempts were made to relate this to the precious metal surface species.

2 Experimental

A general impregnation method was used to prepare all catalysts with a 20 wt % Pt and 10 wt % Ru loading (1:1 mole ratio). Reducing agents and reduction conditions were varied according to Table 1.

Table 1: List of catalyst samples and reduction conditions.

Sample	Reducing agent	Sample	Reducing agent
1	Formaldehyde, 1 h, 10 x excess	7	Formic acid, 1 h, 100 x excess
2	Formaldehyde, 5 h, 10 x excess	8	Formic acid, impregnation in 4 steps
3	Formaldehyde, 1 h, 100 x excess	9	H_2, 923 K
4	Formaldehyde, impregnation in 4 steps	10	H_2, 1 173 K
5	Formic acid, 1 h, 10 x excess	11	No reducing agent, heat treat only
6	Formic acid, 5 h, 10 x excess	JM	Commercial reference

Vulcan XC-72 carbon was suspended in de-ionized water, sodium bicarbonate (10 eq) was added and the mixture was heated to reflux for 30 minutes. A solution of chloroplatinic acid (1 eq) and ruthenium trichloride (1 eq) in de-ionized water was slowly added to the boiling slurry. The mixture was again refluxed and allowed to cool to room temperature. Reduction of the catalyst was performed either in situ by the addition of an appropriate reducing agent, or after filtration under a reducing atmosphere. The catalyst was filtered after it had completely cooled and washed with water until free of chloride ions. The precipitate was dried at 378 K overnight. Finally, the catalyst was heat treated under a nitrogen atmosphere at 623 K for 2 hours. (No additional heat treatment under nitrogen was performed when reduction was performed under a reducing atmosphere.) In the case of catalysts 4 and 8, prepared according to a stepwise procedure, impregnation with only 5 wt % Pt and 2.5 wt % Ru was done according to the above described procedure, and this repeated 4 times in order to achieve a final catalyst loading of 20 wt % Pt and 10 wt % Ru.

High resolution transmission electron microscopy (HRTEM) analyses of all samples were obtained by the Centre for Electron Microscopy at the University of KwaZulu-Natal, with a Jeol 100CX TEM system.

X-ray diffraction (XRD) patterns of the PtRu/C nanocatalysts were obtained with a Siemens Powder Diffraction system (Model D500) using a Cu Kα source operated at 40 keV, 30 mA at a step size of $0.02°$ $2θ$ and a counting time of 3 second per step applied over a range of 5 to $80°$ $2θ$.

Electron probe micro analysis (EPMA) were performed on selected samples by surface mounting small samples of catalyst material on a brass stub, pressed down firmly to compact the material as densely as possible. They were then placed in the Cameca SX50 electron microprobe and compared by measuring against calibrated reference standards of pure Ru, Pt and C. An accelerating voltage of 20 kV and a beam of 20 nA was used as standard settings, with a probe diameter of 5 μm.

Temperature programmed reduction (TPR) was performed on a Micromeritics TPR/TPD 2920 apparatus equipped with a thermal conductivity detector (TCD). Typically, ~200 mg of sample was heated to 573 K under argon, and cooled to 173 K. Samples were heated from 173 K to 373 K at 2 K/min, then to 1 173 K at 10 K/min, under a flow of 10 % H_2/Ar gas mixture.

Electrochemical tests were performed on an Autolab 302N potentiostat equipped with a rotating disk electrode (RDE) at 293 K. A three-electrode system was used in a 0.5 M H_2SO_4 solution, with the addition of 1.0 M methanol for methanol oxidation experiments. Potentials were measured against a saturated calomel reference electrode (SCE). The 5th scan in each run was used to allow measurements to stabilize. Platinum foil was used as counter electrode, with a glassy carbon working electrode with surface area - 0.0707 cm^2 (diameter of 3 mm). The working electrode was polished to a mirror finish with 0.05 μm alumina suspension before each experiment. The catalyst ink was prepared by adding 10 mg of sample to 5 ml de-ionized water and ultrasonicating for 10 minutes. Then 10 μl was pipetted onto the surface of the electrode and allowed to air dry. When completely dry, 10 μl of a 1 wt % Nafion solution was pipetted onto the ink surface. Before each experiment, the solution was de-aerated with Ar for 30 minutes.

All experiments were also carried out on a commercially available 20 wt % Pt, 10 wt % Ru on carbon catalyst (Johnson Matthey) as a reference material.

3 Results and discussion

TEM photographs of catalysts 7 and 10 are shown in Figure 1. Size distributions were obtained by measuring sizes of 50 randomly selected particles in the TEM images. Average particle sizes are listed in Table 2. In most cases well dispersed spherical particles were formed with a small size.

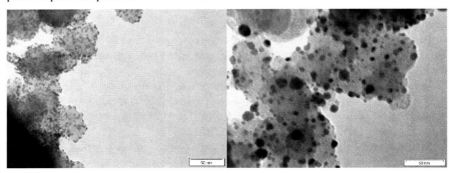

Fig. 1. TEM images of catalysts 7 (left), prepared by reduction with a mild reducing agent - formaldehyde, and 10 (right), reduced under harsh conditions at 900 °C under H_2.

The typical particle size for PtRu/C catalysts is reportedly around 3nm. [9] It has also been reported that a mild reducing agent, like formaldehyde and formic acid, leads to smaller particles. [10] This was found to be the case in this study as well, with catalysts 1-3, reduced by formaldehyde, possessing very small average particle size (2.50 – 2.68 nm). In the case of catalysts 4 and 8, prepared by a stepwise procedure as described in literature, slightly larger metal particles were formed (3.24 and 3.09 nm). Contrary to what was reported, the stepwise impregnation method did not lead to better dispersion of the catalyst on the support or a reduction in crystallite size. [11] Reduction with H_2 was also found to produce small, well dispersed particles (catalyst 9). However when this was done at a higher temperature of 1 173 K (catalyst 10), sintering and agglomeration occurred which leads to an increase in particle size.

Table 2: Average particle size as determined by TEM and XRD.

	Particle size			Particle size	
Sample	TEM	XRD	Sample	TEM	XRD
1	2.69	5.5	7	2.61	4.9
2	2.61	5.8	8	3.09	8.6
3	2.50	5.2	9	2.87	15.6
4	3.24	5.0	10	4.98	40.5
5	2.86	4.8	11	2.47	n/a
6	3.02	5.9	JM	3.02	4.7

XRD patterns for all catalysts are shown in Figure 2. Catalyst samples *1-8* displayed XRD patterns very similar to that of *JM* displaying only Pt peaks, with a slight shift to higher 2θ values. [12] No Ru signals were observed, indicating alloy formation between Pt and Ru. For most catalysts, including *JM*, some sharp peaks are observed between 25 and 35 degrees, as well as 50 and 60 degrees 2θ. These can possibly be attributed to trace amounts of carbonate species originating from $NaHCO_3$ used in the synthesis. EPMA studies also confirm the absence of other contaminating species.

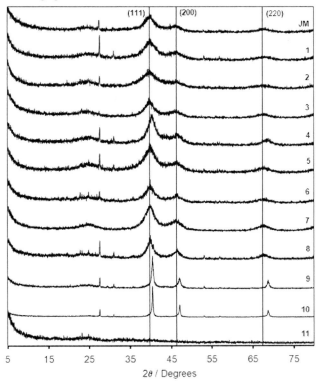

Fig. 2. XRD patterns of all catalysts, including the commercial standard *JM*.

Sharpened peaks were observed for catalysts *9* and *10*, treated at higher temperatures, indicating increased particle size and a more crystalline phase. Catalysts *9* and *10* also showed a more pronounced shift to higher 2θ values. This is due only to a larger degree of alloying taking place during treatment at very high temperatures, and cannot be attributed to a different chemical composition as shown by EPMA studies. [4] Catalyst *11* did not display any XRD peaks, indicating no crystalline metallic phase since no reduction treatment was performed on the sample. Particle sizes were determined from the (200) Pt peaks, using the Scherrer equation. Correlation between TEM and XRD results seems poor, however no standardization was used and results are reported uncorrected. Values do, however, follow the same trends as those obtained from TEM results. A further discrepancy between XRD and TEM results was observed for catalysts *9* and *10*. This indicates that sintering and agglomeration does not take place homogeneously, since XRD is a comprehensive technique yielding an average value for the complete sample, while TEM images of selected areas does not reflect this.

Analysis by electron microprobe (EPMA) was performed on selected catalysts (*1*, *3*, *5*, *9*, *10* and *JM*), and results are listed in Table 3. This is not considered an absolute quantitative technique in this case, since factors like air space between the carbon particles, imperfect sample presentation, etc can play a role. It does however confirm qualitatively the presence of only platinum, ruthenium, carbon and oxygen in the catalysts. Thus confirming that catalysts *9* and *10* do not contain a different chemical composition. Results obtained for the *JM* catalyst does correspond very closely to the supplier's certificate of analysis (19.63 % Pt, 9.7% Ru).

Table 3: Average precious metal content as determined by EPMA.

Sample	Pt (wt %)	Ru (wt %)	Sample	Pt (wt %)	Ru (wt %)
1	24.7	9.6	9	20.4	9.0
3	34.3	6.0	10	18.9	5.7
5	32.3	8.4	JM	19.7	9.7

TPR spectra of catalysts *7*, *11* and *JM* are shown in Figure 3 and the results listed in Table 4. Catalysts *1-3* and *5-8* all displayed a single reduction peak with small intensity at approximately 290 K, similar to that shown for catalyst *7*. Catalysts *4*, *9*, *10* and the *JM* sample all displayed a reduction peak close to 250 K, although the *JM* peak was significantly more pronounced. Catalyst 11, which did not receive any reduction treatment showed a very large peak at 224 K. All catalysts also showed a large reduction peak above 600 K, which can be attributed to the support as previously reported by Huang et al. [13]

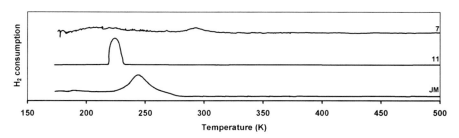

Fig. 3. TPR spectra of catalysts *7*, *11* and the commercially available standard *JM*.

The studies of Huang et al. have shown that pure Pt on carbon catalysts displayed reduction of PtO_x between 230 and 250 K. It was also shown that pure Ru on carbon showed reduction peaks of amorphous RuO_2 and crystalline RuO_2 at 360 and 450 K respectively. For PtRu on carbon the reduction of an alloy oxide species (AO_x) could be observed at 290 K, although the temperature of the reduction peak is known to vary according to the composition of surface metal species. The more PtO_x surface species, the closer the peak to 230 K and the more RuO_x surface species, the closer the peak to 370 K. It is also postulated that the metal particles could possess a core-shell structure with Pt on the surface and Ru in the core (also referred to as a cherry-like structure), thus influencing the reduction properties of the overall catalyst. [7, 11] Similar results were also shown by Jiang et al., who made use of TPR as the reduction step in preparing PtRu on carbon catalysts, as well as Gómez de la Fuente et al. [4, 8, 12]

Table 4: TPR peak temperatures and electrochemical data of all catalysts.

Sample	Temp (K)	Electroactive surface area (m^2/g Pt)	Current density at 0.5 V (mA/mg)	Onset potential (V)
1	296	46	45	0.39
2	299	101	99	0.44
3	287	119	104	0.45
4	245	87	131	0.34
5	287	78	65	0.39
6	291	103	87	0.41
7	293	138	157	0.39
8	284	52	71	0.37
9	248	69	39	0.37
10	253	101	73	0.40
11	224	64	58	0.42
JM	244	96	114	0.36

Relating this information to the present study indicates that the surface of the precious metal particles of catalysts *1-3* and *5-8*, (all reduced in situ by a liquid reducing agent) consists of a mixture of platinum and ruthenium and some alloyed metal oxidized. Catalysts *4*, *9*, *10* and *JM* contain a surface more rich in PtO_x species, possibly indicating a core-shell structure, with the JM catalyst containing more oxide species. Catalyst *11* (which had no reduction treatment) contains a large amount of Pt-rich oxide species on the surface.

The electrochemistry of all samples was investigated in 0.5 M H_2SO_4, and the methanol oxidation properties tested in 1.0 M methanol and 0.5 M H_2SO_4. Electrochemical results of selected samples are shown in Figure 4 and summarized in Table 4. Cyclic voltammograms (CVs) in 0.5 M H_2SO_4 are used to estimate the electrochemically active surface area of the catalyst by using the hydrogen desorption charge, assuming a monolayer hydrogen adsorption charge of 210 $\mu C/cm^2$. This technique is however considered somewhat inaccurate due to the formation of ruthenium oxides on the PtRu alloy surface, but can be used in a comparative way. [13-15]

CVs in methanol showed a methanol oxidation peak at 0.8 to 1.0 V on the forward scan as well as a peak on the reverse scan between 0.4 and 0.6 V. This is reportedly due to the removal of incompletely oxidized carbonaceous species on the surface formed during the forward scan. [16]

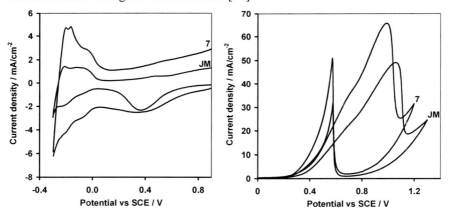

Fig. 4. Cyclic Voltammograms of catalyst *7* and the *JM* standard in (left) 0.5 M H_2SO_4 recorded at 20 mV/s and (right) 1.0 M CH_3OH and 0.5 M H_2SO_4 recorded at 20 mV/s. All spectra were recorded at 20 °C.

It can be seen from Table 4 that similar trends are observed for the electroactive surface area and current density values for methanol oxidation. Reduction of the catalyst is clearly a vital step since the unreduced sample *11* shows poor electrochemical performance. The conditions of reduction are also crucial since reduction with only 10 times excess reducing agent for 1 hour also leads to poor performance (catalysts *1* and *5*). Increasing the amount of reducing agent and the reduction time to 5 hours both lead to catalysts with better electrochemical performance than the commercial standard *JM*. However, best results are obtained when keeping the reduction time short with a large excess of reducing agent (catalysts *3* and *7*). Increased reaction time can lead to unwanted agglomeration, since it is generally reported that smaller particles yields a better catalyst. [9] Reduction with a hydrogen gas flow is only effective when performed at a very high temperature (1 173 K). At this temperature, however, sintering occurs leading to large particles. The catalyst still performs worse than catalysts prepared by milder reducing agents in the liquid phase and cannot be improved further by increasing temperature.

4 Conclusion

The reduction of precious metal particles is a crucial step in the synthesis of preparing Pt, Ru on carbon catalysts. This study has shown that various factors influence the success of such a catalyst. Optimization of the reduction of the metal salts to the metallic state is crucial. Incomplete reduction leads to poor electrochemical performance, although some surface oxide species is still needed for oxidation of CO to CO_2 as suggested by the bifunctional mechanism described earlier. A particle size of approximately 3 nm is considered standard, although it is generally accepted that the smaller the particle size, the better the catalyst. Reduction under sub-optimum conditions can lead to agglomeration or sintering, leading to larger metal particles and a less effective catalyst. It was found that a short reduction time with a large excess of reducing agent in the liquid phase produces a better catalyst, even compared to the commercial standard.

Acknowledgements

We thank Ms Sharon Eggers at the Centre for Electron Microscopy at the University of KwaZulu-Natal for performing TEM observations. This work was funded by Mintek and the Department of Science and Technology as part of the HySA/Catalysis project.

References

1 Kamarudin S.K, Achmad F, Daud W.R.W (2009) Overview on the application of direct methanol fuel cell (DMFC) for portable electronic devices. Int J Hydrogen Energy, 34: 6902-6916.
2 Guo J.W, Zhao T.S, et al (2005) Preparation and characterization of a PtRu/C nanocatalyst for direct methanol fuel cells. Electrochim Acta, 51: 754-763.
3 Long J.W, Stroud R.M, et al (2000) How to make electrocatalysts more active for direct methanol oxidation-Avoid PtRu bimetallic alloys! J Phys Chem B, 104: 9772-9776.

4 Gómez de la Fuente J.L, Martínez-Huerta M.V, et al (2009) Tailoring and structure of PtRu nanoparticles supported on functionalized carbon for DMFC applications: New evidence of the hydrous ruthenium oxide phase. Appl Catal B: Enivronmental, 88: 505-514.

5 Ma L, Liu C, et al (2009) High activity PtRu/C catalysts synthesized by a modified impregnation method for methanol electro-oxidation. Electrochim Acta, 54: 7274-7279.

6 Samra L.S, Lin T.D, et al (2005) Carbon-supported Pt-Ru catalysts prepared by the nafion stabilized alcohol-reduction method for application in direct methanol fuel cells. J Power Sources, 139: 44-54.

7 Huang S, Chang S, et al (2006) Promotion of the electrochemical activity of a bimetallic platinum-ruthenium catalyst by oxidation-induced segregation. J Phys Chem B, 110: 23300-23305.

8 Gómez de la Fuente J.L, Pérez-Alonso F.J, et al (2009) Identification of Ru phases in PtRu based electrocatalysts and relevance in the methanol electrooxidation reaction. Catal Today, 143: 69-75.

9 Wang K, Huang S, et al (2007) Promotion of cabon-supported platinum-ruthenium catalyst for electrodecomposition of methanol. J Phys Chem C, 111: 5096-5100.

10 Tian J.H, Wang F.B, et al (2004) Effect of preparation conditions of Pt/C catalysts on oxygen electrode performance in proton exchange membrane fuel cells. J Appl Electrochem, 34: 461-467.

11 Kinoshita K, Stonehart P (1977) Preparation and Characterization of Highly Dispersed Electrocatalytic Materials. In: Bockris J.O.M, Conway B.E, Modern aspects of electrochemistry, vol. 12, Chapter 4 Plenum Press, New York, 183-266.

12 Wang K, Yeh C (2008) Temperature-programmed reduction study on carbon-supported platinum-gold alloy catalysts. J Colloid & Interface Sci, 325: 203-206.

13 Huang S, Chang S, et al (2006) Characterization of surface composition of platinum and ruthenium nanoalloys dispersed on active carbon. J Phys Chem B, 110: 234-239.

14 Jiang L, Sun G, et al (2005) Preparation of supported PtRu/C electrocatalyst for direct methanol fuel cells. Electrochim Acta, 50: 2371-2376.

15 Jiang J, Kucernak A (2003) Electrooxidation of small organic molecules on mesoporous precious metal catalysts II: CO and methanol on platinum-ruthenium alloy. J Electroanal Chem, 543: 187-199.

16 Cao D, Bergens S.H (2003) A nonelectrochemical reductive deposition of ruthenium adatoms onto platinum: anode catalysts for a series of direct methanol fuel cells. Electrochim Acta, 48: 4021-4031.

17 Kinoshita K, Ross P.N (1977) Oxide stability and chemisorption properties of supported ruthenium electrocatalysts. J Electroanal Chem, 78: 313-318.

18 Lin M, Lo M, et al (2009) PtRu nanoparticles supported on ozone-treated mesoporous carbon thin film as highly active anode materials for direct methanol fuel cells. J Phys Chem C, 113: 16158-16168.

Synthesis and investigation of silver-peptide bioconjugates and investigation in their antimicrobial activity[1]

O. Yu. Golubeva*, O. V. Shamova**, D. S. Orlov.**, E. V. Yamshchikova**, A. S. Boldina*, V. N. Kokryakov**

* Institute of Silicate Chemistry of the Russian Academy of Sciences, St.-Petersburg, Russia

** Institute for Experimental Medicine of the Russian Academy of Medical Sciences, St.-Petersburg, Russia

golubeva@isc.nw.ru, olga_isc@mail.ru

Abstract

Bioconjugates on the basis of silver nanoparticles and antimicrobial proline-rich peptide C-Bac3.4 were synthesized. The biological activity of the bioconjugates synthesized was tested in comparison with activity of their constituents – nanoparticles and the peptide. It was found that the nanoparticles-peptide conjugates exhibited the pronounced antimicrobial activity against bacteria including that resist to conventional antibiotics – drug-resistant clinical isolate of *Pseudomonas aeruginosa* and the methicillin resistant strain of *Staphylococcus aureus*. It was shown that bioconjugates studied did not exhibit the pronounced membrano-lytic activity typical of most of antimicrobial peptides. The results obtained let us make the conclusion that the bioconjugates synthesized are characterized by the properties which are different from the properties of their constituents – the antimicrobial peptide and silver nanoparticles.

Key words: silver nanoparticles, antimicrobial peptides, bioconjugates, antibiotics, antimicrobial activity

Introduction

The combination of natural and synthetic components into a hybrid entity is one of the fundamental principles of bioconjugation. Use of these materials in drug delivery is well established [1-3]. The purpose of our research was to extent the application of conjugated materials and to combine inorganic nanoparticles

---

[1] The title was extended according to the evaluators advice.

and biological molecules with aim to develop new materials for medical applications.

It is well-known that silver ions and silver-based compounds are highly toxic to microorganisms [4, 5] showing strong biocidal effects on as many as 16 species of bacteria. Silver nanoparticles should be of greatest interest for research because it is known that substances in the nano state exhibit unusual properties, which are not observed at the macro- and micro-state. In addition, the silver nanoparticles are more stable and can longer maintain their biological activity than the ions. It has been found recently [6] that the biocidal effect of silver nanoparticles greatly exceeds the effect of silver ions Ag^+ at the same concentrations.

Antimicrobial peptides are the effector molecules of the innate immune system [7]. They are widely distributed in nature and have been isolated from a variety of sources including bacteria, invertebrates, vertebrates and plants [8, 9]. Antimicrobial peptides exhibit a broad spectrum of killing activity against various targets, such as bacteria, fungi, enveloped viruses, parasites and tumour cells [10, 11]. However, research articles devoted to finding concrete ways to implement these peptides in clinical practice are still rare, despite the obvious benefits of such drugs: they have no antigenic properties, as well as the ability of microorganisms to develop resistance to these peptides is very low.

The main aim of the present research was to obtain complexes of inorganic nanoparticles and natural antibiotic molecules to compare the biological action of such complexes with that of the original structures.

Experimental

Materials

As the biological component of the synthesized bioconjugates the proline-rich antimicrobial peptide C-Bac3.4 (CRFRLPFRRPPIRIHPPPFYPPFRPFL) was chosen. This peptide is a modification of a natural peptide ChBac3.4 isolated from leukocytes of the goat; C-Bac3.4 has an additional amino acid cysteine at the N-terminus of the molecule, for more effective conjugation with metal nanoparticles.

Isolation, structural characterization and antimicrobial properties of ChBac3.4 have been previously described [7].

Stock solution of 0.1 M $AgNO_3$ (Aldrich) and 0.3 mM peptide were dissolved in doubly deionized water.

Bioconjugates synthesis

Synthesis technique is modified from the previously described in literature procedure [12] which was used for obtaining of bioconjugates with short HRE peptides. The peptide stock solution (0.7 mL) was added to 0.1 mL of $AgNO_3$ solution and 5 mL of 0.1 M Tris-HCl buffer (pH 8.6). After 15 min of stirring 0.1 mL of 0.1 M $NaBH_4$ was added dropwise. The reduction was carried out for 4 h at which point the silver clusters were completely formed.

Nanoparticles conjugated with the peptide were separated from the unbound peptide by centrifugation at $40000 \times g$ for two hours. The precipitate obtained was resuspended in phosphate buffered saline containing150 mM of NaCl, pH 7.4.

Product characterization

Protein concentration in bioconjugates synthesized was examined by Bradford assay [13]. Nanoparticles in aqueous solution were studied by UV-visible spectrophotometer (LEKI SS2109UV, SpectraMax250) and Transmission Electron Microscopy (EM-125 electron microscope, U_{accel}=75 kV).

The bioconjugates synthesized were tested for antimicrobial activity against *Escherichia coli* ML35p, *Pseudomonas aeruginosa* ATCC 27853, *Pseudomonas aeruginosa*, *Staphylococcus aureus* SG511, *Listeria monocytogenes* EGD by using the standard technique of microdillution broth assay [14]. The overnight cultures of each strain in Mueller-Hinton Broth were transferred to a fresh media and feather incubated to obtain a mid-logarithmic phase of bacteria. Absorbance of bacterial suspensions was measured at 620 nm; then the suspensions were diluted to approximately 2 x 10^5 CFU/ml and mixed with bioconjugates dilutions in wells of a microplate. After incubation for 18 h at 37°C the minimal inhibitory concentration (MIC) was read as a highest dilution of the bioconjugate sample resulting in the complete inhibition of visible growth of microorganisms.

To examine the ability of samples synthesized to permeabilize the inner membranes of gram-negative bacteria *Escherichia coli*, a previously described procedure [15] was used. The indicator strain allows us to monitor changes in the permeability of an inner membrane. If under the influence of a destructive agent bacterial membranes become permeable to certain marker molecules, the optical density at a wavelength of 420 nm increases, which allows us to observe the process of damaging the inner membrane of the bacteria in real time.

Results and discussion

The precipitate obtained after the centrifugation divided into two fractions - grayish - black precipitate (I), forming large flakes, and yellowish-brown (II), well suspended in buffer. Yellow color of the supernatant is symptomatic of the presence of silver nanoparticles in the solution.

According to the results of the Bradford assay the precipitates contained protein. This fact confirms the formation of complexes between silver nanoparticles and peptides.

Fig. 1 shows the UV-vis spectra of all fractions, obtained after the centrifugation. For all fractions the maximum of the absorption was observer in a range of 400-450 nm. This characteristic resonance responds to excitation of surface plasmon vibrations in the silver nanoparticles and is responsible for striking colors of the samples.

Fig. 2 represents the TEM micrographs of the sample before centrifugation. As can be seen from Fig. 2, the bi-phased sample was obtained which coincide with the spectroscopy data. There are some silver nanoparticles with sizes around 5 nm. The aggregations with an average particles size ranging from 50 to 150 nm are also observed. It can be assumed that these aggregations are probably conjugates of silver nanoparticles, encapsulated in the peptide environment. The presence of silver and peptide in the precipitate are confirmed by optical spectroscopy and the results of determination of protein by Bradford assay.

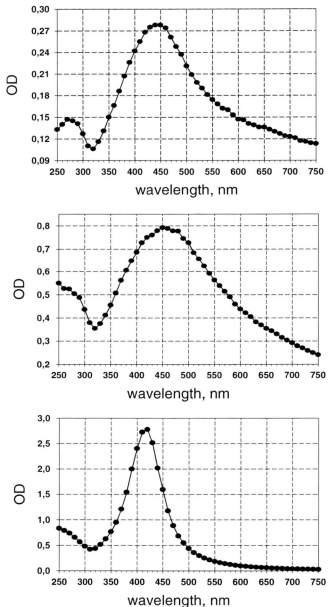

Fig. 1: UV-vis spectra from precipitate I (a) and precipitate II (b) redispersed in a buffer, and supernatant (c):

Synthesis and investigation of silver-peptide bioconjugates

Fig. 2: TEM micrographs of silver nanoparticles-antimicrobial peptide bioconjugates

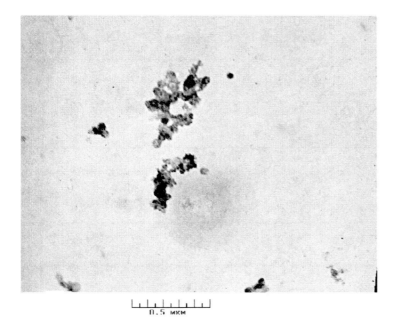

Fig. 3: TEM micrographs of precipitate II

Comparison of the results obtained by electron microscopy and optical spectroscopy, as well as the results of the determination of protein by Bradford assay, allows us to conclude that the precipitates I and II contain silver nanopaticles-antimicrobial peptide bioconjugates. The conjugates core probably has a fractal structure formed by silver nanoparticles. This is evidenced by the strong long-wave broadening of the plasmon absorption observed in optical spectra of the samples [16], and electron microscopy data presented in Fig . 3.

Comparison of the results obtained by electron microscopy and optical spectroscopy, as well as the results of the determination of protein by Bradford assay, allows us to conclude that the precipitates I and II contain silver nanopaticles-antimicrobial peptide bioconjugates. The conjugates core probably has a fractal structure formed by silver nanoparticles. This is evidenced by the strong long-wave broadening of the plasmon absorption observed in optical spectra of the samples [16], and electron microscopy data presented in Fig . 3.

Results of the antimicrobial activity testing are shown in Table 1.

Table 1. Minimal inhibitory concentration (MIC) of different fractions of the silver-peptide bioconjugates obtained by centrifugation. MIC is represented as a highest dilution of the sample resulting in the complete inhibition of visible growth of microorganisms

Microorganisms	MIC		
	Precipitate I	Precipitate II	Supernatant
Escherichia coli ML35p	1/2	1	1/4
Pseudomonas aeruginosa ATCC 27853	1/2	1/2	1/4
Pseudomonas aeruginosa clinical isolate, resistant to antibiotics	1/2	1/2	1/4
Staphylococcus aureus SG511	1/4	1/2	1/4
Listeria monocytogenes EGD	1/2	1/2	1/2
MRSA ATCC 33591	1/2	1/2	1/2

It is followed from the Table 1, that silver-peptide bioconjugates synthesized are highly active against different microorganisms including a resistant to conventional antibiotics strain of *Pseudomonas aeruginosa*, and methicillin-resistant *Staphylococcus aureus* strain. It should be noted that we have earlier identified the antimicrobial activity of free silver nanoparticles [17], but in higher concentrations than tested in this study. The results obtained allow us to conclude that the silver-peptide bioconjugates are characterized by the broad-spectrum antibiotic activity exceeded activity of their constituents – the antimicrobial peptide and silver nanoparticles.

To compare the mechanisms of antimicrobial action of the conjugate synthesized with its constituents – silver nanoparticles and the peptide, their effect on the permeability of the *Escherichia coli ML35p* inner membrane was studied. It should be noted that the functional and structural integrity of the inner membrane is one of the most important conditions of the bacterium survival. The destruction

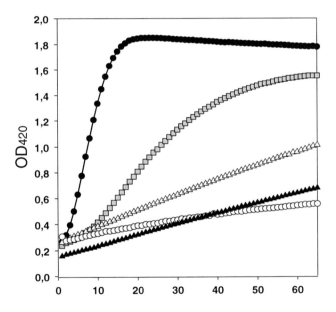

Fig. 4. Membrane permeabilization. We tested the ability of bioconjugates synthesized to permeabilize the inner membranes of Escherichia coli ML35p and compared its action with effects of antimicrobial peptide C-Bac3.4, silver nanoparticles and a membranolytic agent protegrin. The extent of membrane permeabilization is proportional to the slope of each curve, and the time of maximal permeabilization.
● - protegrin, □ - C-Bac3.4, △ - silver nanoparticles, ○ – bioconjugates, ▲ – control

of the bacterium inner membrane in particular can be considered as the reliable indicator of the bactericide action [18]. The results demonstrating the influence of the substances studied on the permeability of the *Escherichia coli ML35p* inner membrane are presented on Fig. 4. For comparison the results for a well-studied membrano-lityc peptide protegrin are also presented. It was established that adding of the C-Bac3.4 to the bacteria at the peptide concentrations higher than MIC leads to the disintegration of the inner bacterial membrane.

In the presence of silver nanoparticles the insignificant increase of the membrane permeability was observed. Silver-peptide conjugates do not demonstrate any effect on the membrane barrier function since they do not cause the increase of the *Escherichia coli* membrane permeability. Therefore the antibiotic action of silver-peptide bioconjugate differs from the antimicrobial peptide one and is not connected with the membrano-lytic mechanism. As far as the membrane destruction is the basis of the antimicrobial peptides toxic action the conjugation with silver nanoparticles can be considered as one of the possible ways to optimize the antimicrobial peptides action.

Concluding remarks

Conjugation of the antimicrobial peptide C-Bac3.4 and silver nanoparticles allows us to obtain the hybrid substances with qualitatively new properties. Bioconjugates obtained have a significantly higher antimicrobial activity and the ability to suppress growth of antibiotic-resistant microorganisms. Conjugation of antimicrobial peptides and silver nanoparticles can be regarded as a way to get the antibiotic substances with new properties, promising to overcome the resistance of microorganisms.

Acknowledgments

This work was supported by Russian Foundation for Basic Research (09-03-12192 ofi_m)

References

1. Tkachenko A, Xie H, Coleman D et al (2003) Multifunctional gold nanoparticle-peptide complexes for nuclear targeting. J Am Chem Soc 125:4700-4701

2. Sokolova V, Radte I, Heumann R et al (2006) Effective transfection of cells with multi-shell calcium phosphate-DNA nanoparticles. Biomaterials 27:3147-3153

3. Berry V, Rangaswamy S, Saraf RF (2004) Highly selective, electrically conductive monolayere of nanoparticles on live bacteria. Nano Lett 4: 939-942

4. Slawson RM, VanDyke MI, Lee H., Trevors JT (1992) Germanium and silver resistance, accumulation, and toxicity in microorganisms. Plasmid 27:72-79

5. Zhao J, Stevens SE (1998) Multiple parameters for the comprehensive evaluation of the susceptibility of Escherichia coli to the silver ion. BioMetals 11: 27-32

6. Morones JR, Elechiguerra JL, Camacho A et al (2005) The bactericidal effect of silver nanoparticles. Nanotechnology 10:2346-2353

7. Shamova O, Orlov D, Stegeman C et al (2009) ChBac3.4: A novel proline-rich antimicrobial peptide from goat leukocytes. Int J Rept Res Ther 15:31-47

8. Hancock RF, Diamond G (2000) The role of cationic antimicrobial peptides in innate host defenses. Trends Microbiol 8:402-410

9. Hancock RE, Sahl HG (2006) Antimicrobial and host-defense peptides as new anti-infective therapeutic strategies. Nat Biotechnol 24:1551-1557

10. Pag U, Oedenkoven M, Sass V, Shai Y, Shamova O et al (2008) Analysis of in vitro activities and modes of action of synthetic antimicrobial peptides derived from an α-helical "sequence template". J Antimicrobial Chemotherapy 61:341-352

11. Lehrer RI, Ganz T (1999) Antimicrobial peptides in mammalian and insect host defence. Curr Opin Immunol 11: 23-27

12. Slocik JM, Wright DW (2003) Biomimetic mineralization of nobel metal nanoclusters. Biomacromolecules 4:1135-1141

13. Bradford M (1976) A Rapid and Sensitive Method for the Quantitation of Microgram Quantities of Protein Utilizing the Principle of Protein-Dye Binding. Anal Biochem 72:248-254

14. Tossi A, Scocchi M, Zanetti M, Genaro R et al (1997) An approach combining cDNA amplification and chemical synthesis for the identification of novel, cathelicidin-derived, antimicrobial peptides. In: Shafer W (ed) Antibacterial peptide protocols, Shpringer, Humana Press

15. Lehrer RI, Barton A, Ganz T (1988) Concurrent assessment of inner and outer membrane permeabilization and bacteriolysis in E.coli by multiple-wavelength spectrophotometry. J Immunol Methods 108:153-158

16. Shalaev VM (1996) Electromagnetic properties of small-particle composites Phys Rep 272:61-137

17. Golubeva OYu, Shamova OV, Orlov DS et al (2010) Investigation of antimicrobial and hemolityc activity of silver nanoparticles obtained by chemical reduction method. Glass Phys Chem (in press)

18. Zhao C, Nguyen T, Boo LM et al (2001) RL-37, an alpha-helical antimicrobial peptide of the rhesus monkey. Antimicrob Agents Chemother 45:2695-2702

Characterization of Stabilized Zero Valent Iron Nanoparticles

Lauren F. Greenlee[1], Stephanie Hooker[2]

Materials Reliability Division, National Institute of Standards and Technology
325 Broadway
Mail Stop 853
National Institute of Standards and Technology
Boulder, CO 80305
USA

[1]Corresponding author (Email: lauren.greenlee@nist.gov, phone: +1-303-497-4234, fax: +1-303-497-5030)

[2]Stephanie Hooker: stephanie.hooker@nist.gov

1   Abstract

The demonstrated toxicity of certain groups of organic micropollutants in water sources has motivated research in developing novel materials that are able to remove dissolved organic molecules from an aqueous system through adsorption and/or degradation. One approach is to use the enhanced surface properties of nano-sized particles to adsorb, reduce, or oxidize organic contaminants. Our research focuses on the use of catalytic nanoparticles to degrade haloamides, a specific family of disinfection by-products (DBPs) produced during chlorine and chloramine disinfection. This work focuses on the development and characterization of zero valent iron-based catalytic nanoparticles. In particular, different stabilizer compounds are used during nanoparticle synthesis to control the particle size and prevent aggregation. The size, shape, and functional groups of the stabilizer compounds are investigated; the roles of specific chelating groups, such as phosphates and carboxylates, in controlling particle size are compared. Particles are characterized through several techniques including dynamic light scattering, electron microscopy, and measurement of zeta potential.

2   Key Words

Carboxymethyl cellulose; phosphonates ; nanoparticles; zero valent iron; stabilization

3   Introduction

Characterization of nanoparticle systems is necessary for effective design of particle-based water treatment technology. Today, catalytic nanoparticles based on metals such as zero valent iron (ZVI) are being considered as potential materials

---

[*]Contribution of NIST, an agency of the US government; not subject to copyright in the United States.

for the adsorption and degradation of a variety of water contaminants. In an aqueous environment, ZVI is oxidized by either molecular oxygen or water, and these reactions cause the oxidation (Keenan and Sedlak 2008) or reduction (Sayles et al. 1997), respectively, of organic compounds. Oxidation by oxygen produces strong radical oxidants such as the hydroxyl radical ($^{\bullet}$OH), while oxidation by water produces molecular hydrogen, which is used in reductive dehalogenation reactions.

The ability of ZVI nanoparticles to reduce halogenated organic compounds makes these particles an attractive treatment system for both ground water remediation and drinking water treatment (Li et al. 2010; Zhang 2003). Research in ground water remediation has shown that ZVI nanoparticles can be successfully injected into a contaminated site and degrade water contaminants (Elliott and Zhang 2001), while ZVI nanoparticles have more recently been considered for drinking water treatment due to the identification of many dissolved organic water contaminants (Richardson et al. 2007; Webb et al. 2003). One group of contaminants, the halogenated disinfection by-products (DBPs), is produced during chlorine-based disinfection processes and is a likely candidate for ZVI nanoparticle degradation. These micropollutants, although present in water sources at concentrations as low as nanograms per liter, are recalcitrant compounds that often pass through water treatment systems into finished drinking water. Furthermore, many DBPs have been shown to be both cytotoxic and genotoxic (Plewa et al. 2008; Plewa et al. 2004; Zwiener et al. 2007).

While ZVI is also reactive at larger length scales, nanosized ZVI particles degrade contaminants more rapidly (Wang and Zhang 1997), which is likely due to the large increase in the surface area to volume ratio as the particle size decreases. A review of literature revealed there might be a nanoparticle size cutoff (40 – 50 nm), above which nano-sized particles behave as micron-sized particles (Auffan et al. 2009). To make ZVI nanoparticles a viable treatment system, the particle size must be controlled and particle aggregation prevented. Recent work has shown that various nontoxic organic polymers can be used to stabilize ZVI nanoparticles during particle synthesis, enabling decreased particle aggregation and increased reactivity (Doong and Lai 2005; He and Zhao 2005; Lin et al. 2009), as well as control of particle size (He and Zhao 2007). Many of the polymers tested contain carboxyl functional groups that coordinate with iron in solution, but organophosphorus compounds, which contain phosphate functional groups, have been shown to more strongly coordinate with di- and trivalent cations such as iron (Jonasson et al. 1996; Yang et al. 2001).

The objective of this study was to compare the ability of several organic compounds to control ZVI nanoparticle size and aggregation. A cellulose-based polymer with carboxylic acid functional groups was compared to three different organophosphorus compounds. ZVI nanoparticles were characterized by electron microscopy, dynamic light scattering, and measurement of zeta potential.

4 Materials and Methods

4.1 Chemicals[1]

All chemicals used were ACS reagent grade. Ferrous sulfate ($FeSO_4*7H_2O$), sodium borohydride ($NaBH_4$), and carboxymethyl cellulose (CMC) were used as received. The CMC had a molecular weight of 250 kg/mol and a degree of substitution of 0.7. Amino tri(methylene phosphonic acid) (ATMP), diethylenetriamine penta(methylene phosphonic acid) (DTPMP), and bis(hexamethylene triamine penta(methylenephosphonic acid)) (HTPMP) were obtained from Dequest Water Management Additives by Thermphos (Anniston, AL). All solutions were made with purified and deionized water.

4.2 Preparation of ZVI Nanoparticles

The nanoparticle synthesis method was based on the work of He and Zhao (2005). All synthesis reactions were performed in a glass round-bottom flask, and mixing was achieved with an orbital shaker at 100 rpm. For unstabilized nanoparticles, an aqueous solution of 1 g/L iron (5.0 g/L $FeSO_4*7H_2O$) was bubbled with argon for 10 min. Particle synthesis was initiated by adding $NaBH_4$ dropwise while mixing by hand. Due to the competing reaction between BH_4^- and water, $NaBH_4$ was added in slight (10 %) excess of the stoichiometric requirement (BH_4:Fe molar ratio of 2:1) to ensure complete reaction of the iron. After $NaBH_4$ addition, the solution was mixed under vacuum until the reaction finished (indicated by the cessation of $H_{2(g)}$ evolution). The particles produced by the reaction are assumed to be primarily zero valent iron, based on previous work (Lee et al. 2009; Lin, et al. 2009; Zheng et al. 2009). Excess dissolved salt was removed by centrifuging the particle solution, removing the supernatant, and replacing the supernatant with purified and deionized water (two washes performed). Samples were dried in a vacuum oven (-75,000 Pa) at room temperature. For stabilized particles, the same procedure was used, except that the initial solution of 1 g/L iron included the desired amount of stabilizer (CMC, ATMP, DTPMP, or HTPMP). CMC was tested at a ratio of 0.0005 mole CMC:1 mole Fe, and the three phosphonate compounds were tested at a ratio of 0.05 mole phosphonate:1 mole Fe.

4.3 Characterization Techniques

Particle size measurements, zeta potential measurements, and pH titrations were performed on a Zetasizer Nano ZS (Malvern Instruments, Westborough, MA) using noninvasive backscatter (NIBS) detection at an angle of 173°. This instrument uses Mie theory to predict the intensity of the light scattered from the particles; the use of Mie theory rather than the Fraunhofer approximation allows

---

[1] Commercial equipment, instruments, or materials are identified only in order to adequately specify certain procedures. In no case does such identification imply recommendation or endorsement by the National Institute of Standards and Technology, nor does it imply that the products identified are necessarily the best available for the purpose.

more accurate measurement of particle size in the nanometer range. At least 10 separate measurements were taken on each sample for both particle size and zeta potential; particle size and zeta potential were measured as point measurements directly following particle synthesis without pH adjustment. The pH was measured following each synthesis, and all solutions had pH values between 8.5 and 9.0. All titrations were performed on aqueous samples starting at pH 2 and increasing to pH 12 with 0.1×10^{-3} mol/m^3 and 0.025×10^{-3} mol/m^3 NaOH. Samples were measured directly following synthesis without the washing or drying procedures mentioned above. Samples were diluted 1:20 with water and filtered with a 0.2 µm polyethersulfone syringe filter before measurement.

Particles were also characterized with field emission scanning electron microscopy (FESEM) and transmission electron microscopy (TEM). Samples were dropcast onto sample holders, vacuum dried, carefully rinsed with water to remove excess salt, and vacuum dried. FESEM samples were prepared on either carbon stubs or silicon wafers. TEM samples were prepared on copper grids. Particle composition was analyzed on the FESEM by energy dispersive x-ray spectroscopy.

5   Results and Discussion

The effect of the stabilizer compounds on nanoparticle size and aggregation was analyzed using dynamic light scattering. Representative results for nanoparticle solutions containing stabilizer are shown in Figure 1 for both particle scattering intensity and particle volume. All four stabilizers resulted in bimodal distributions of the scattering intensity (Figure 1a). While all samples displayed bimodal scattering intensity distributions, the volume distribution results indicate that most of the particle volume was contained in the smaller (10-20 nm) size population, except for the HTPMP sample. The addition of CMC, ATMP, or DTPMP stabilizers to the solution before nanoparticle synthesis resulted in a population of small particles with diameters between 5 nm and 50 nm and a population of larger particles between 50 nm and 1,000 nm. However, the presence of the HTPMP stabilizer during nanoparticle synthesis resulted in larger populations of approximately 100 nm and 1.5 µm. The mode at 100 nm represents individual particles that formed during synthesis, as can be seen in FESEM images in Figure 2b-d. However, the larger mode likely represents agglomerated particles because particles larger than 100 nm to 200 nm were not observed. These results were repeatable from batch to batch of synthesized nanoparticles. In comparison, an unstabilized nanoparticle sample (image shown in Figure 2a) resulted in both unimodal and bimodal distributions over a series of 10 consecutive measurements. The bimodal distribution of the unstabilized particles was similar to the bimodal distribution shown for HTPMP-stabilized nanoparticles in Figure 1a (data represented by open circles). The average particle diameter (calculated by instrument software) increased from approximately 400 nm to 800 nm during the 10 measurements (20 minute time lapse).

The average modal particle diameter of the smaller population was similar for CMC (11.0 nm +/- 5.7 nm) and ATMP (11.7 nm +/- 0.8 nm) samples, and slightly larger for DTPMP (21.7 nm +/- 2.5 nm). However, the modal particle diameter measurements of the smaller population for the CMC sample varied greatly, while ATMP samples had the smallest standard deviation. The large variation in measurements for the CMC sample may result from the presence of CMC itself, which is a much larger molecule that any of the phosphonates. The interactions of CMC with the nanoparticles and the amount of CMC adsorbed to particle surfaces likely changes the measured particle diameter. Dynamic light scattering measures the hydrodynamic diameter of particles, and the presence of any stabilizer at the surface of the nanoparticles might increase the measured particle diameter. The polydispersity index, a measurement of the mass distribution, also varied between the four stabilizers. HTPMP had the lowest average index (0.43 +/- 0.08), while ATMP had the highest average index (0.96 +/- 0.07). DTPMP had an average polydispersity index of 0.69 +/- 0.07 and CMC had an average index of 0.48 +/- 0.06. Note that similar small particle populations were obtained for the CMC, ATMP, and DTPMP samples, but the mass of the phosphonate (ATMP and DTPMP) necessary to produce this size distribution was smaller than that of CMC by a factor of 10.

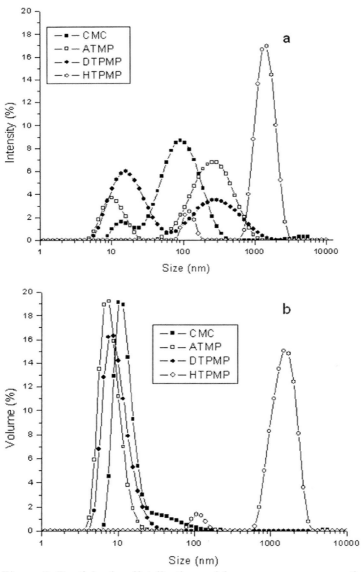

Figure 1. Particle size distributions of four aqueous nanoparticle suspensions with each of the four stabilizers obtained by dynamic light scattering measurement, shown as (a) scattering intensity and (b) volume. All samples contained 1 g/L Fe.

The conversion of the scattering intensity data to a volume distribution (Figure 1b) slightly decreased the modal particle diameter of the peaks, which is due to the deconvolution of the scattering intensity data and not to a change in the actual particle size. However, the volume distribution data are useful in determining the sample composition of different particle size populations. The samples containing ATMP and DTPMP as stabilizers resulted in unimodal distributions with only a peak at the smaller particle diameter. The sample containing CMC resulted in a similar peak. However, the CMC volume distribution curve has a shoulder on the right side of the curve, indicating some particle volume in larger particle sizes. The HTPMP sample resulted in a bimodal distribution similar to that of the scattering intensity distribution (Figure 1a). Dynamic light scattering results indicate that both carboxyl- and phosphate-functionalized stabilizer compounds can be used to produce nanoparticles with diameters in the range of 5nm to 20 nm. All of the stabilized samples resulted in consistent particle size measurements and showed no signs of aggregation over time, as was the case for the unstabilized sample. Within all of the stabilized samples, at least two size groups of particles were formed. This result could potentially lower the effectiveness of the particles for contaminant degradation.

Electron microscopy was used to identify the particle size populations measured by dynamic light scattering; FESEM results are shown in Figure 2. The unstabilized sample (Figure 2a) contained a wide distribution of particle sizes, with most individual particles having diameters less than 300 nm. However, most of the particles were formed in chains, and few separate particles were observed. The chains and larger aggregates observed were most likely formed immediately following synthesis because the sample formed visible aggregates even before the synthesis reaction went to completion. The particle aggregation during synthesis was most likely responsible for the variable particle size distributions obtained by dynamic light scattering. As mentioned above, the dynamic light scattering measurements taken on unstabilized nanoparticle suspensions resulted in both unimodal and bimodal particle size distributions. The unimodal particle size distributions did not show a mode at 100 nm, even though nanoparticles of this size were observed with electron microscopy. Therefore, the particles were aggregated and measured as aggregates during dynamic light scattering.

In contrast, all the stabilized samples resulted in individual, dispersed particles, and the stabilizer compound is clearly visible around and between the nanoparticles (Figure 2b-d). All the stabilized samples appear to have relatively monodisperse particle populations, and the larger particle size population identified by dynamic light scattering for ATMP, DTPMP, and CMC is similar to the size of the particles visible in the FESEM images. This result indicates that the difference in the polydispersity indices of the stabilized samples is likely due to the presence of the stabilizer and is not a result of the particles themselves. No particle chains were observed when stabilizer was added to the sample prior to ferrous iron reduction and nanoparticle formation. Sample preparation and amount of sample deposited on the sample substrate most likely caused the high density of particles shown in Figure 2b-d. The samples were centrifuged to separate the particles from excess stabilizer and dissolved salts, redispersed in methanol, and deposited on silica wafers for FESEM analysis. The evaporation of the methanol caused the nanoparticle sample to visibly dry in uneven patterns. Thus, the images were taken from areas with a high density of particles. Dynamic light scattering data of redispersed stabilized samples indicate that centrifuged samples, including the samples shown in Figure 2b-d, are not irreversibly aggregated.

EDX analysis on the FESEM samples confirmed the presence of iron particles and little to no residual sodium or sulfate; the particles contained a small amount of oxygen, indicating some oxidation of the particles might have occurred. However, when compared to oxidized particles, the oxygen content (weight %) of a fresh sample was only one third to one half that of an oxidized sample. EDX analysis of stabilized samples indicated that these samples contained larger amounts of carbon and oxygen, likely due to the stabilizer compounds, and the presence of phosphorus was confirmed in the phosphonate-stabilized samples.

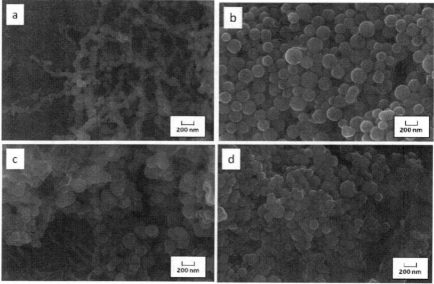

Figure 2. FESEM images of (a) unstabilized ZVI nanoparticles and particles stabilized by (b) CMC, (c) ATMP, and (d) DTPMP. Unstabilized particles aggregate and form chains, while stabilized particles are dispersed within stabilizer, and no chains are observed.

TEM allowed identification of the smaller particle population initially identified through dynamic light scattering, with particles between 5 nm and 20 nm observed (Figure 3). Particles smaller than 10 nm in diameter were observed for both samples, and the unstabilized sample also contained aggregated particles. The larger size populations observed in FESEM images were also identified in TEM. Many small particles (<10 nm) were observed in the stabilized sample (Figure 3b), but due to the presence of stabilizer, the sample was diluted to obtain clear images of a smaller number of particles, resulting in a small number of particles within one image. As in the larger particle population observed in the FESEM images, the unstabilized sample contained a wider distribution of particle sizes than that of the CMC-stabilized sample. Furthermore, the stabilized particles appeared to be more uniform in shape and more dispersed, with no large aggregates or particle clumping observed. All small particles observed in the CMC-stabilized sample were less than 10 nm, while the modal particle diameter obtained by dynamic light scattering was 11.0 nm +/- 5.7 nm, as mentioned above. As previously suggested, the discrepancy in observed particle size is likely the result of CMC adsorbed to nanoparticle surfaces and larger measured hydrodynamic diameters. Diffraction patterns were able to be obtained for the unstabilized sample, while no diffraction patterns were able to be observed in the CMC-stabilized sample; this result indicates that although the nanoparticles in the unstabilized sample are crystalline, the presence of stabilizer might allow at least the small nanoparticles (less than 20 nm in diameter) to remain amorphous during the precipitation and crystallization process. As particles nucleate and grow, the initial particle phases are often unstable and amorphous (Clarkson et al. 1992; Munemoto and Fukushi 2008). Without the presence of stabilizer, the nucleated particles form ordered crystal structures as the particles grow. However, previous work has shown that the presence of additives or stabilizers can both slow or arrest the particle growth phase and allow the particles to remain in an amorphous phase (Greenlee et al. 2010).

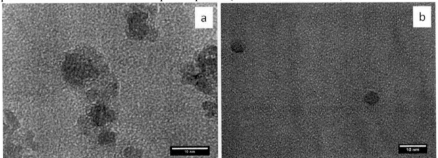

Figure 3. TEM images of (a) unstabilized ZVI nanoparticles and (b) particles stabilized by CMC. Scale bars are 10 nm.

The zeta potential of each nanoparticle sample was measured from pH 2 to pH 12. Results for the CMC, ATMP, and unstabilized samples are shown in Figure 4. The zeta potential for the CMC-stabilized sample was negative throughout the entire pH range tested, while the ATMP-sample had zeta potentials similar to those of the unstabilized sample. Results obtained for DTPMP and HTPMP were similar to those of the ATMP sample. The relationship between zeta potential and pH was similar for the unstabilized sample and the sample stabilized with ATMP. Both samples had positive zeta potential values for pH values less than 8 or 9; the ATMP-stabilized sample had an isoelectric point (IEP) at 7.8, while the unstabilized sample had an IEP of 8.9. The DTPMP sample had an IEP of 6.9, and the HTPMP sample had an IEP of 8.0 (data not shown). In contrast, the CMC-stabilized sample had a negative zeta potential over the entire range of tested pH values. At pH values above 6, the ATMP-stabilized sample had zeta potential values that were slightly lower than those of the unstabilized sample; in a basic environment, the zeta potential of the ATMP-stabilized particles approached the zeta potential of the CMC-stabilized particles.

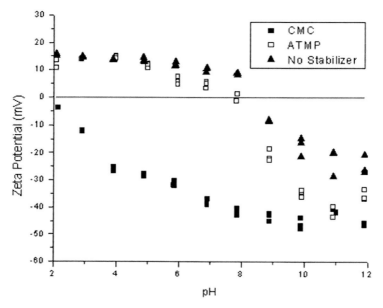

Figure 4. Nanoparticle zeta potential as a function of pH for an unstabilized sample and for samples stabilized by CMC and ATMP. The three data points at each pH for each sample represent three separate experiments.

These results indicate that the charge of the particles was significantly influenced by the CMC polymer, while the phosphonate stabilizers had a much smaller effect. The size of the molecules and how the stabilizers are coordinated around the nanoparticles, as well as the different pH values, or protonation constant (pKa) values, when the different functional groups dissociate, could also contribute to the differences in zeta potential over the pH range. The pKa values of the amine groups of phosphonates are typically above 9 (Popov et al. 2001); therefore, most of the amines would be protonated at pH values below 9. The carboxyl groups of carboxymethyl cellulose have pKa values below pH 4, which indicates that the carboxyl groups would be deprotonated over most of the pH range tested (Heinze and Koschella 2005). Aminophosphonate compounds, such as those used in this study, have phosphate pKa values in the range of 4 to 8; therefore, some of the phosphate groups would be deprotonated but some would remain protonated over the acidic to neutral pH range (Popov, et al. 2001). This knowledge of pKa values suggests that in the CMC sample, most of the carboxyl groups were deprotonated over the pH range tested. However, in the phosphonate samples, the amine groups were protonated, and only some of the phosphate groups were deprotonated, until above pH 9. This difference in protonation could also have influenced the different zeta potential results obtained.

The influence of zeta potential and pH on average particle size was also investigated for the pH range 2 to 12; results for samples containing DTPMP and HTPMP are shown in Figure 5. The sample stabilized by DTPMP appeared to have a pH-dependent average particle size, while the HTPMP-stabilized sample had an average particle size that increased during the pH titration. Note that the average particle size diameters start at pH 2 as values indicative of the larger particle size populations described in Figure 1a; the average particle size is calculated from the scattering intensity data and therefore is skewed towards the larger particle sizes. Nanoparticle samples containing DTPMP or ATMP (data not shown) displayed a correlation between zeta potential and average particle size. Samples containing CMC, HTPMP, or no stabilizer showed no relationship between zeta potential and average particle size, even though the CMC sample had a zeta potential trend distinctly different than from those of the HTPMP or unstabilized samples. For the DTPMP and ATMP samples, the average particle size increased dramatically as the positive zeta potential approached the IEP, and the average particle size then decreased as the zeta potential became increasingly negative. For the other three sample types, the particle sizes either remained relatively stable, as the case for unstabilized particles, or increased at approximately pH 6 (CMC and HTPMP).

The increase in average particle size observed for CMC and HTPMP samples might have resulted from irreversible particle aggregation or particle oxidation over the time period of the titration measurements (three to four hours). In a separate study, the particle size of a CMC sample was measured over several days, and the particle size increased from the initial measurement. Previous work suggests that stabilizers might prevent or delay particles oxidation (Geng et al. 2009) but our handling of aqueous samples has shown that stabilized aqueous nanoparticles do eventually oxidize, albeit more slowly than unstabilized samples, over several weeks. However, during the autotitration from pH 2 to 12, the sample is mixed and recirculated through the tubing and sample holders, which could promote both particle aggregation and particle oxidation. Interestingly, the ATMP and DTPMP stabilizers appear to prevent this trend of increased particles size. Although particles appear to aggregate at pH values around the IEP of the sample, the particles appear to redisperse in solution as the zeta potential moves away from the IEP.

Titration results indicate that the zeta potential and, in some cases, the particle size could be controlled or chosen based on the choice of stabilizer compound and working pH. The ability to control sample parameters such as particle size and particle charge could become important when optimizing the particles for a treatment system. The varying results between the three phosphonate-stabilized samples show that even similar molecular structures can have significantly different effects on particle size and particle stabilization. Furthermore, while all four stabilizers tested resulted in small, stabilized particles, there may be a critical working time for particle stabilization for some stabilizer compounds.

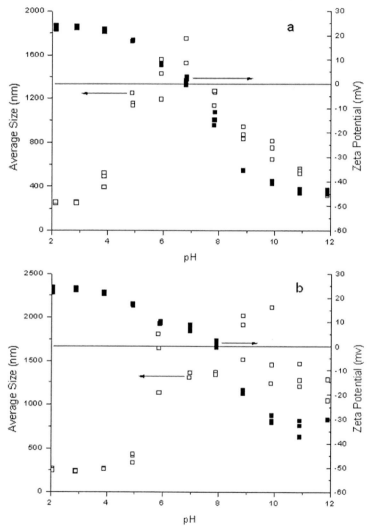

Figure 5. Comparison of size and zeta potential measurements as a function of pH for nanoparticle samples stabilized by (a) DTPMP and (b) HTPMP. The three data points at each pH for each sample represent three separate experiments.

6 Conclusions

While nano-sized particles are produced without particle stabilization, the addition of organic stabilizer before particle precipitation prevented particle aggregation and particle chain formation. All four stabilizers produced bimodal particle size distributions, with varying modal particle diameters and polydispersity indices. ATMP- and CMC-stabilized samples resulted in the smallest modal particle diameter. For three of the four stabilizers tested, a particle population within 5 – 20 nm was observed, and this mode represented the majority of the particle volume. HTPMP did not cause the formation of particles less than 50 nm. Two phosphonate stabilizers, ATMP and DTPMP, at concentrations 10 fold less than CMC, produced particles with modal diameters similar to those stabilized by CMC. The presence of CMC resulted in a negative zeta potential over the pH range 2 – 12, while phosphonate-stabilized samples had slightly basic isoelectric points, similar to the unstabilized sample. ATMP- and DTPMP-stabilized nanoparticles had a pH-dependent average particle size, while the other stabilized and unstabilized samples had average particle sizes that increased over time during pH titrations. These results indicate that some phosphonate compounds, such as ATMP and DTPMP, can be used, at lower concentrations than CMC, to produce nanoparticles, and the zeta potential and average particle size could potentially be controlled using pH. However, greater control over particle size, a unimodal particle size distribution, and lower stabilizer concentrations are desirable. Further research might consider other stabilizer compounds and variations on experimental conditions to improve these aspects of nanoparticle synthesis.

7 Acknowledgements

The authors acknowledge Roy H. Geiss for his work obtaining the TEM images, as well as Ann N. Chiaramonti, Robert R. Keller, and Nicholas Barbosa III for their help with FESEM imaging and EDX analysis.

8 References

Auffan M, Rose J, Bottero JY et al (2009) Towards a definition of inorganic nanoparticles from an environmental, health and safety perspective. Nature Nanotechnology 4(10):634-641 10.1038/nnano.2009.242

Clarkson JR, Price TJ and Adams CJ (1992) Role of metastable phases in the spontaneous precipitation of calcium carbonate. Journal of the Chemical Society-Faraday Transactions 88(2):243-249

Doong RA and Lai YJ (2005) Dechlorination of tetrachloroethylene by palladized iron in the presence of humic acid. Water Research 39(11):2309-2318 10.1016/j.watres.2005.04.036

Elliott DW and Zhang WX (2001) Field assessment of nanoscale biometallic particles for groundwater treatment. Environmental Science & Technology 35(24):4922-4926 10.1021/es0108584

Geng B, Jin ZH, Li TL et al (2009) Preparation of chitosan-stabilized Fe-0 nanoparticles for removal of hexavalent chromium in water. Science of the Total Environment 407(18):4994-5000 10.1016/j.scitotenv.2009.05.051

Greenlee LF, Testa F, Lawler DF et al (2010) Effect of antiscalants on precipitation of an RO concentrate: Metals precipitated and particle characteristics for several water compositions. Water Research 44(8):2672-2684 doi:10.1016/j.watres.2010.01.034

He F and Zhao DY (2005) Preparation and characterization of a new class of starch-stabilized bimetallic nanoparticles for degradation of chlorinated hydrocarbons in water. Environmental Science & Technology 39(9):3314-3320 10.1021/es048743y

He F and Zhao DY (2007) Manipulating the size and dispersibility of zerovalent iron nanoparticles by use of carboxymethyl cellulose stabilizers. Environmental Science & Technology 41(17):6216-6221 10.1021/es0705543

Heinze T and Koschella A (2005) Carboxymethyl ethers of cellulose and starch - A review. Macromolecular Symposia 223:13-39 10.1002/masy.200550502

Jonasson RG, Rispler K, Wiwchar B et al (1996) Effect of phosphonate inhibitors on calcite nucleation kinetics as a function of temperature using light scattering in an autoclave. Chemical Geology 132(1-4):215-225

Keenan CR and Sedlak DL (2008) Factors affecting the yield of oxidants from the reaction of manoparticulate zero-valent iron and oxygen. Environmental Science & Technology 42(4):1262-1267

Lee YC, Kim CW, Lee JY et al (2009) Characterization of nanoscale zero valent iron modified by nonionic surfactant for trichloroethylene removal in the presence of humic acid: A research note. Desalination and Water Treatment 10(1-3):33-38

Li TY, Chen YM, Wan PY et al (2010) Chemical Degradation of Drinking Water Disinfection Byproducts by Millimeter-Sized Particles of Iron-Silicon and Magnesium-Aluminum Alloys. Journal of the American Chemical Society 132(8):2500-2501 10.1021/ja908821d

Lin YH, Tseng HH, Wey MY et al (2009) Characteristics, morphology, and stabilization mechanism of PAA250K-stabilized bimetal nanoparticles. Colloids and Surfaces a-Physicochemical and Engineering Aspects 349(1-3):137-144 10.1016/j.colsurfa.2009.08.007

Munemoto T and Fukushi K (2008) Transformation kinetics of monohydrocalcite to aragonite in aqueous solutions. Journal of Mineralogical and Petrological Sciences 103(5):345-349

Plewa MJ, Muellner MG, Richardson SD et al (2008) Occurrence, synthesis, and mammalian cell cytotoxicity and genotoxicity of haloacetamides: An emerging class of nitrogenous drinking water disinfection byproducts. Environmental Science & Technology 42(3):955-961

Plewa MJ, Wagner ED, Jazwierska P et al (2004) Halonitromethane drinking water disinfection byproducts: Chemical characterization and mammalian cell cytotoxicity and genotoxicity. Environmental Science & Technology 38(1):62-68

Popov K, Ronkkomaki H and Lajunen LHJ (2001) Critical evaluation of stability constants of phosphonic acids (IUPAC technical report). Pure and Applied Chemistry 73(10):1641-1677

Richardson SD, Plewa MJ, Wagner ED et al (2007) Occurrence, genotoxicity, and carcinogenicity of regulated and emerging disinfection by-products in drinking water: A review and roadmap for research. Mutation Research/Reviews in Mutation Research 636(1-3):178-242

Sayles GD, You GR, Wang MX et al (1997) DDT, DDD, and DDE dechlorination by zero-valent iron. Environmental Science & Technology 31(12):3448-3454

Wang CB and Zhang WX (1997) Synthesizing nanoscale iron particles for rapid and complete dechlorination of TCE and PCBs. Environmental Science & Technology 31(7):2154-2156

Webb S, Ternes T, Gibert M et al (2003) Indirect human exposure to pharmaceuticals via drinking water. Toxicology Letters 142(3):157-167 10.1016/s0378-4274(03)00071-7

Yang QF, Liu YQ, Gu AH et al (2001) Investigation of calcium carbonate scaling inhibition and scale morphology by AFM. Journal of Colloid and Interface Science 240(2):608-621

Zhang WX (2003) Nanoscale iron particles for environmental remediation: An overview. Journal of Nanoparticle Research 5(3-4):323-332

Zheng ZH, Yuan SH, Liu Y et al (2009) Reductive dechlorination of hexachlorobenzene by Cu/Fe bimetal in the presence of nonionic surfactant. Journal of Hazardous Materials 170(2-3):895-901 10.1016/j.jhazmat.2009.05.052

Zwiener C, Richardson SD, De Marini DM et al (2007) Drowning in disinfection byproducts? Assessing swimming pool water. Environmental Science & Technology 41(2):363-372

Combustion Synthesis of Nanoparticles CeO$_2$ and Ce$_{0.9}$Gd$_{0.1}$O$_{1.95}$

Sumittra Charojrochkul[1]*, Waraporn Nualpaeng[2,4], Navadol Laosiripojana[2], Suttichai Assabumrungrat[3]

1 National Metal and Materials Technology Center, Thailand Science Park, Pathumthani, Thailand

2 The Joint Graduate School in Energy and Environment, Bangkok, Thailand

3 Dept. of Chemical Engineering, Chulalongkorn University, Bangkok, Thailand

4 National Nanotechnology Center, Thailand Science Park, Pathumthani, Thailand

Abstract Ceria and Gd$_2$O$_3$ doped ceria nanoparticle powder has been fabricated using a combustion synthesis method for an application as an electrolyte for Intermediate Temperaure Solid Oxide Fuel Cell (ITSOFC). The nanoparticle powder is obtained when the processing parameters are optimized. Three different fuels i.e. urea, citric acid and glycine have been used for producing the nanoparticle powder of doped ceria. The produced particles have been characterized using SEM for the particle morphology, XRD for phase identification, and particle analysis for the particle size distribution and identification. In addition, these particles have shown good reforming activity, which provide an opportunity for this material not to be used only as an electrolyte but also as a support for the anode part of SOFC.

1 Introduction

The world energy crisis in the 1990s of fossil fuel shortages, climate change and widespread pollution have led to overwhelming research in alternative energy. One of the most important hydrogen conversion tools is a fuel cell, especially Solid Oxide Fuel Cells (SOFCs), which show advantages concerning a fuel efficiency, emissions, maintenance and noise pollution. A fuel cell is interesting for alternative energy since it offers a long-term potential for a highly efficient energy system and is based on available resources.

Theoretically, the major fuel for SOFCs system is hydrogen-rich gas (the mixture of H$_2$ and CO), which can be produced by steam reforming, dry reforming or partial oxidation. However, hydrogen rarely exists in its free form naturally. Steam methane reforming (SMR) is the most common and least expensive method in producing hydrogen. Correspondingly, 95% of hydrogen production in the US uses steam reforming technology. In these reactions, catalyst is needed to assist the decomposition of hydrocarbon to H_2/CO-rich gas. Doped ceria has been extensively used as a three-way catalyst in the catalytic converter for automobiles. More recently, this material has been introduced as an electrolyte for an intermediate temperature Solid Oxide Fuel Cell (IT-SOFC) to replace the conventional Yttria Stabilized Zirconia (YSZ) which had been a conventional electrolyte for high temperature SOFC. Dopants such as Gd_2O_3 and Y_2O_3 have a crucial role for improving the ionic conductivity of this electrolyte to be applicable for SOFC at around 550-650°C. Recently, ceria-based catalyst and perovskite-based catalyst (Sauvet AL, Fouletier J 2001; Barison S. et al 2007) have been extensively studied as well as noble metal catalysts for reforming process. Ceria-based materials used as a reforming catalyst has been studied by several researchers (Laosiripojana N, Assabumrungrat S 2006; Pino L et al 2006) due to its good reactivity and better resistance toward carbon deposition than conventional Ni catalysts (Laosiripojana N, Assabumrungrat S 2005). Moreover, the use of ceria had been successful for an SOFC operating directly on dry methane reforming (Laosiripojana N, Assabumrungrat S 2005). In addition, Zr (Fornasiero P et al 1996)-, Gd (Ramirez-Cabrera E et al 2002; Hennings U, Reimert R 2007)- and Nb (Ramirez-Cabrera E et al 2002)-doped ceria have also been investigated to enhance its catalytic activity due to the effect of introducing oxygen vacancies to maintain the overall charge neutrality (Skinner SJ, Kilner JA 2003).

In the present study, ceria was selected as based catalyst and synthesized using a combustion synthesis technique. This preparation technique is considered as a cost effective technique compared to conventional ceramic processing techniques (Patil KC et al 2002). This technique has been proven to be capable of producting nanoparticles of ceria and doped ceria for several applications (Nualpang W et al 2008; Charojrochkul S et al 2008; Chen W et al 2006; Lenka R et al 2008; Hwang C et al 2006; Chinarro E et al 2007; Mahata T et al 2005; Biswas M et al 2007; Xu H et al 2008; Aranda A et al 2009). CeO_2 and $Ce_{0.9}Gd_{0.1}O_{1.95}$ were synthesized using three different fuels i.e. urea, citric acid and glycine for producing nanoparticle powder of doped ceria. The produced particles were characterized using SEM for the particle morphology, XRD for phase identification, and particle analysis for the particle size distribution and identification. The catalytic activity of powder was studied via methane steam reforming reaction.

2 Experimental

2.1 Catalyst preparation

CeO_2 and $Ce_{0.9}Gd_{0.1}O_{1.95}$ (CGO) were prepared using a combustion synthesis technique. Cerium (III) nitrate hexahydrate ($Ce(NO_3)_3 \cdot 6H_2O$, 99.5%) and Gadolinium (III) nitrate hexahydrate ($Gd(NO_3)_3 \cdot 6H_2O$, 99.9%) supplied from Alfa Aesar were used as initial chemical reactants. Urea, glycine and citric acid were used as fuel. A stoichiometry between the fuel and oxidant was calculated by balancing the net reducing valence of the fuel and the net oxidizing valence of the oxidant. The oxidizing valences of the oxidants and the reducing valences of different fuels are given in Table 1. The stoichiometric molar ratios of fuel to metal nitrate or oxidant ratio were determined based on the above concept. The reducing valences of nitrogen and carbon have been taken as 0 and 4, respectively considering N_2 and CO_2 as the gaseous combustion products. Therefore, the stoichiometry between oxidant and urea, glycine and citric acid is 2.5:1, 1.67:1 and 0.83:1, respectively. Combustion reactions of each reaction can be expressed as follows, where ψ is the fuel to oxidant ratio.

CeO_2 prepared from urea-nitrate combustion synthesis

$$Ce(NO_3)_3 \cdot 6H_2O + \psi NH_2CONH_2 + \left(\frac{3}{2}\psi - \frac{7}{2}\right)O_2 \rightarrow CeO_2 + \left(\frac{3}{2} + \psi\right)N_2 + \psi CO_2 + (6 + 2\psi)H_2O \quad (1)$$

$Ce_{0.9}Gd_{0.1}O_{1.95}$ prepared from urea-nitrate combustion synthesis

$$0.9Ce(NO_3)_3 \cdot 6H_2O + 0.1Gd(NO_3)_3 \cdot 6H_2O + \psi NH_2CONH_2 + \left(\frac{3}{2}\psi - \frac{141}{40}\right)O_2 \rightarrow \quad (2)$$
$$Ce_{0.9}Gd_{0.1}O_{1.95} + \left(\frac{3}{2} + \psi\right)N_2 + \psi CO_2 + (6 + 2\psi)H_2O$$

CeO_2 prepared from citric acid-nitrate combustion synthesis

$$Ce(NO_3)_3 \cdot 6H_2O + \frac{5}{6}C_6H_8O_7 + \frac{1}{4}O_2 \rightarrow CeO_2 + \frac{3}{2}N_2 + 5CO_2 + \frac{28}{3}H_2O \quad (3)$$

$Ce_{0.9}Gd_{0.1}O_{1.95}$ prepared from citric acid-nitrate combustion synthesis

$$0.9Ce(NO_3)_3 \cdot 6H_2O + 0.1Gd(NO_3)_3 \cdot 6H_2O + \frac{5}{6}C_2H_8O + \frac{109}{40}O_2 \qquad (4)$$

$$\rightarrow Ce_{0.9}Gd_{0.1}O_{1.95} + \frac{3}{2}N_2 + 5CO_2 + \frac{28}{3}H_2O$$

CeO_2 prepared from glycine-nitrate combustion synthesis

$$Ce(NO_3)_3 \cdot 6H_2O + \frac{5}{3}NH_2CH_2COOH + \frac{1}{4}O_2 \rightarrow CeO_2 + \frac{7}{3}N_2 + \frac{10}{3}CO_2 + \frac{61}{6}H_2O \qquad (5)$$

$Ce_{0.9}Gd_{0.1}O_{1.95}$ prepared from glycine-nitrate combustion synthesis

$$0.9Ce(NO_3)_3 \cdot 6H_2O + 0.1Gd(NO_3)_3 \cdot 6H_2O + \frac{5}{3}NH_2CH_2COOH + \frac{9}{40}O_2 \rightarrow Ce_{0.9}Gd_{0.1}O_{1.95} \qquad (6)$$

$$+ \frac{7}{3}N_2 + \frac{10}{3}CO_2 + \frac{61}{6}H_2O$$

Cerium (III) nitrate hexahydrate ($Ce(NO_3)_3 \cdot 6H_2O$) and fuel were mixed for the preparation of cerium oxide. For the preparation of gadolinium doped ceria (CGO), cerium (III) nitrate hydrate and Gadolinium (III) nitrate hydrate ($Ga(NO_3)_3 \cdot 6H_2O$) were mixed with each fuel as the desired ratio. To obtain a homogenous solution, a minimum volume of deionized water around 5 millilitres were added to the mixed reactant. The crucible containing the mixed reactant was then heated by means of a Bunsen burner with an excess temperature until an autoignition occurred. The as-synthesized powders were pulverized for 10 minutes.

Table 1. The oxidizing valences of the oxidants and the reducing valences of fuels (Lenka RK et al 2008)

Compound	Derivation	Valency oxidizing (−) reducing (+)
Oxidizers		
$M(NO_3)_3$; M=Ce, Gd	3 + 3. (0 + 3.(−2))	−15
Fuels		
Urea (CH_4N_2O)	4+ 4.(1) + 2.(0) + (−2)	6
Citric acid ($C_6H_8O_7$)	6.(4) + 8.(1) + 7.(−2)	18
Glycine($C_2H_5NO_2$)	2.(4) + 5.(1)+0+2.(−2)	9

2.2 Characterization

The X-ray diffractometer (JEOL, Model JDX-3530) using Cu-Kα in the 2θ range from 20° to 100° was used to identify the as-prepared powders for phase identification. The specific surface area of catalyst powder was measured using a nitrogen adsorption isotherm BelsorpII. Particle morphology and microstructure are examined using a scanning electron microscope (JEOL, Model JSM6301F). Raman spectrometry (Bruker – Senterra, laser source at 785 nm) was carried out to identify the surface species.

The activities of catalyst towards the methane steam reforming were determined by loading each catalyst in a stainless steel reactor. An amount of 200 mg of the catalyst was loaded in this reactor and packed with a small amount of quartz wool to prevent the catalyst from dispersing. The feed gases of 5% methane in He flown at the rate of 100 ml/min and steam at the flow rate of 0.8 ml/h were introduced into the reactor. The temperature of the system was increased from room temperature to 900°C at the rate of 10°C/ min. After the reforming reaction, the exit gas mixture was transferred to a mass spectrometer (Quadrupole mass spectrometers QMS, PFEIFFER) for examinations. The gases of interest were H_2, CO, CO_2 and CH_4.

3 Results and Discussion

3.1 Feature of combustion synthesis

The powders prepared using each fuel were hereafter identified as U, G and C for powder prepared from urea, glycine and citric acid, respectively. For example, U-CeO_2 is referred to cerium oxide prepared from urea and G-CGO is identified the gadolinium-doped ceria prepared from glycine. A theoretical approach has been taken to evaluate the exothermicity of combustion reactions. The heat of the combustion reaction has been determined using the relation [1]. The calculated thermodynamic quantities of combustion reactions of each reaction are given in Table 2.

$$\Delta H = \Sigma (n \cdot \Delta H_f)_{product} - \Sigma (n \cdot \Delta H_f)_{reactants} \qquad [1]$$

where n is the number of the mole of product and initial reactants for the first and second terms, respectively. The calculated thermodynamic quantities of combustion reaction of cerium oxide were shown in Table 2. The values indicate that the preparation of cerium oxide using urea as a fuel provides the lowest amount of heat while citric acid fuel provides the highest amount of heat.

Table 2. Calculated thermodynamic quantities of combustion reaction of cerium oxide.

Fuel	Fuel to oxidant ratio	Enthalpy (kJ)
Urea	2.5:1	583.9
Glycine	1.67:1	663.5
Citric Acid	0.83:1	712.1

Figs. 1, 2 and 3 show the combustion reaction of different fuels i.e. urea-, glycine- and citric acid-nitrate, respectively. The combustion reactions of each fuel are rather different. The reaction of urea-combustion synthesis was found to propagate throughout the reactant and the product powder was uniform light yellow. In the citric acid combustion, the reactant form a volume of gel and then the combustion reaction occurred. The powder was non-uniform in colour. Some appear in grey colour. In case of glycine-combustion, the reaction was rather vigorous. When the combustion occurred, the powders were dispersed vigorously and powders obtained were very fine.

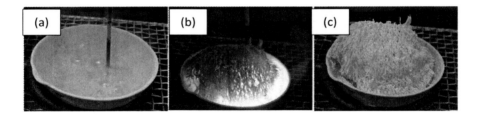

Fig. 1. Combustion reaction of urea-nitrate.

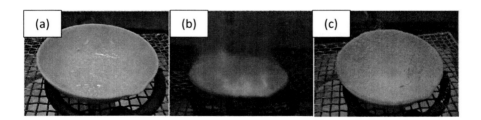

Fig. 2. Combustion reaction of glycine-nitrate.

Fig. 3. Combustion reaction of citric acid-nitrate.

3.2 Physical properties of the catalyst powder

Figs. 4 and 5 show X-ray diffraction patterns of CeO_2 and CGO, respectively. All the synthesized catalysts formed a cubic fluorite phase without any other phases in these patterns. To calculate the crystallite sizes, XRD patterns of (111) plane of powders were scanned at a rate of 0.01°/min. The crystallite sizes were calculated using Scherrer's formula:

$$D = \frac{0.9\lambda}{\beta \cos\theta} \qquad [2]$$

where λ is the X-ray wavelength (0.154060 nm), β is the corrected width at half peak of the maximum intensity (radian), θ is the diffraction angle. β can be calculated from:

$$\beta^2 = \beta_m^2 - \beta_s^2 \qquad [3]$$

where β_m is the measured FWHM, β_s is the FWHM of a standard sample, The value of β_s was taken as 0.117. The crystallite sizes were reported in Table 3.

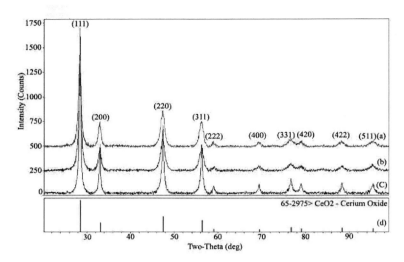

Fig. 4. X-ray diffraction patterns of CeO_2 (a) U-CeO_2, (b) G-CeO_2, (c) C-CeO_2 and (d) CeO_2 standard JCPDS 65-2975.

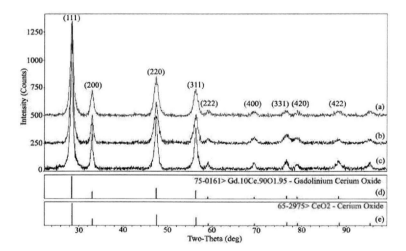

Fig. 5. X-ray diffraction patterns of CGO (a) U-CGO, (b) G-CGO, (c) C-CGO, (d) $Gd_{0.1}Ce_{0.9}O_{1.95}$ standard JCPDS 75-0161 and (e) CeO_2 standard JCPDS 65-2975.

Table 3 reported the effect of the type of fuel on surface area and crystallite size of the synthesized powders. The surface area of powder prepared from citric acid is the largest compared with those prepared from urea and glycine.

Table 3. Particle size, surface area and crystallite size of powder prepared from different fuels.

Sample code	Surface area (m^2/g)	Crystallite size (nm)
U-CeO$_2$	51.0	11.36
G-CeO$_2$	12.4	10.93
C-CeO$_2$	59.6	11.82
U-CGO	22.8	10.97
G-CGO	27.9	20.92
C-CGO	59.6	12.88

To study the morphology and microstructure of synthesized powders, the scanning electron microscope was used to capture impages at a magnification of X20,000. Figs. 6 and 7 depict the morphology of CeO$_2$ and CGO, respectively, when prepared from different fuels. The powders of each condition were smaller than 500 nanometers. The particles prepared from urea combustion appeared to be the largest in size where the particles prepared from glycine and citric acid were finer. The size of particles of all the fuel combustion has corresponded well with the amount of heat released from the exothermic reactions shown in Table 2.

Fig. 6. SEM micrographs of CeO$_2$ (a) U-CeO$_2$, (b) G-CeO$_2$ and (c) C-CeO$_2$.

Fig. 7. SEM micrographs of pulverized Ce$_{0.9}$Gd$_{0.1}$O$_{1.95}$ powder a) U-CGO, (b) G-CGO and (c) C-CGO.

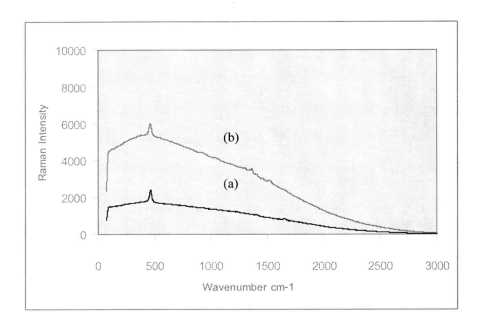

Fig. 8. Raman spectra of (a) CeO2 and (b) CGO particles.

The formation of solid solution of CGO was confirmed with FT-Raman spectra in Fig.8 that both spectra for CeO$_2$ and CGO were detected very close together at around wave number of 460-464 /cm. The results were similar to those reported earlier by Godinho ML et al 2007.

3.3 Catalytic property toward methane steam reforming

Table 4 shows the catalytic properties of the synthesized catalysts. The synthesized powders were studied over the methane steam reforming at 900°C. The reforming rate was measured as a function of time to investigate the stability and the deactivation rate. The conversion of methane in steam reforming at 900°C over CeO$_2$ and CGO prepared from different fuels are shown in Fig. 9 and Fig. 10, respectively. It was found that the methane conversions for all samples were in the same range and not directly related to the specific surface area. Nevertheless, it can be noticed that the catalyst prepared with urea showed relatively higher stability than others.

The catalytic properties of CeO$_2$ (Riedel-deHean), CeO$_2$ (Nanophase Technology Corporation) and 10 mol% gadolinia-doped ceria (Rhodia) were also studied to compare with the prepared catalysts. CeO$_2$ from Nanophase Technology Corporation is a nano-particle size powder. CeO$_2$ (Riedel-deHean), CeO$_2$ (Nanophase Technology Corporation) and 10 mol% gadolinia-doped ceria (Rhodia) were hereafter called R-CeO$_2$, nano-CeO$_2$ and CGO10K. It was found that the methane conversions of synthesized powders were in the same length as the commercial catalyst.

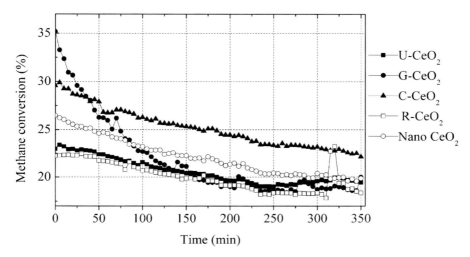

Fig. 9. Stream reforming of methane for the synthesized CeO$_2$

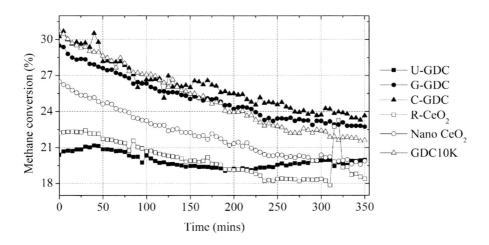

Fig. 10. Steam reforming of methane for the synthesized CGO

In this study, the ratio of carbon monoxide to the summation of carbon monoxide and carbon dioxide (CO/CO+CO$_2$) were used to analyze the degree of water gas shift reaction in the system, from which the lower ratio indicates a higher water gas shift reaction. There was no sample that showed the ratio greater than commercial catalysts. Therefore, catalysts prepared by combustion synthesis technique had higher ability to change carbon monoxide and steam to hydrogen via the water gas shift reaction better than commercial catalysts. The synthesized powders thus tend to provide higher hydrogen than commercial catalysts.

Table 4. %CH$_4$ conversion at steady state, deactivation and the ratio of carbon monoxide to the summation carbon monoxide and carbon dioxide.

Sample Code	%CH$_4$ conversion at steady state	Deactivation (%)	$\dfrac{CO}{CO + CO_2}$
U-CeO$_2$	19.5	15.3	0.165
G-CeO$_2$	20.0	43.2	0.156
C-CeO$_2$	22.2	25.1	0.142
U-CGO	19.9	2.2	0.179
G-CGO	22.7	22.9	0.140
C-CGO	23.7	22.8	0.164
CeO$_2$-R	18.4	17.2	0.190
Nano-Ceria	19.9	24.9	0.234
CGO10K	21.6	29.6	0.178

4 Conclusions

Nanoparticles of CeO$_2$ and Ce$_{0.9}$Gd$_{0.1}$O$_{1.95}$ were successfully prepared via combustion synthesis technique by using urea, glycine and citric acid as fuel. All samples show a single solid solution of cubic fluorite structure. Regarding its methane steam reforming activity, the methane conversions of all synthesized powders were in the same range and close to that of the commercial catalysts. Moreover, the synthesized catalyst was found to provide higher ability to convert carbon monoxide and steam to hydrogen via the water gas shift reaction than that of the commercial catalysts. CGO has a high potential to be a practical anode for IT-SOFC.

Acknowledgments

The authors would like to acknowledge research funding from National Research Council of Thailand (NRCT) and the Joint Graduate School of Energy and Environment (JGSEE) at King Mongkut's University of Technology Thonburi with research support from National Metal and Materials Technology Center. W. Anuchitolarn is thankful for Raman investigation.

5 References

Aranda A, López J et al (2009) Total oxidation of naphthalene with high selectivity using a ceria catalyst prepared by a combustion method employing ethylene. *Journal of Hazardous Materials* 171(1-3) 393-399.

Barison S, Battagliarin M et al (2007) Novel $Au/La_{1-x}Sr_xMnO_3$ and $Au/La_{1-x}Sr_xCrO_3$ composites: Catalytic activity for propane partial oxidation and reforming. *Solid State Ionics* 177(39-40): 3473-3484.

Biswas M, Prabhakaran et al (2007) Synthesis of nanocrystalline yttria doped ceria powder by urea-formaldehyde polymer gel auto-combustion process. *Materials Research Bulletin* 42(4) 609-617.

Charojrochkul S, Nualpang W et al (2008) Combustion Synthesis of Doped Ceria Multioxide for Use as an Electrolyte for SOFCs, the 8th European Solid Oxide Fuel Cell Forum, 30 June – 4 July, 2008, Lucerne, Switzerland.

Chen W, Li F et al (2006) A facile and novel route to high surface are ceria-based nanopowders by salt-assisted solution combustion synthesis. *Materials Science and Engineering:B* 133 (1-3) 151-156.

Chinarro E, Jurado J et al (2007) Synthesis of ceria-based electrolyte nanometric powders by urea-combustion technique. *Journal of the European Ceramic Society* 27(13-15) 3619-3623.

Fornasiero P, Balducci G et al (1996) Modification of the redox behaviour of CeO_2 induced by structural doping with ZrO_2. *Journal of Catalysis* 164(1) 173-183

Godinho MJ, Gonçalves RF et al (2007) Room temperature co-precipitation of nanocrystalline CeO_2 and $Ce_{0.8}Gd_{0.2}O_{1.9-\delta}$ powder. *Materials Letters* 61: 1904-1907

Hennings U, Reimert R (2007) Investigation of the structure and the redox behavior of gadolinium doped ceria to select a suitable composition for use as catalyst support in the steam reforming of natural gas. *Applied Catalysis A: General* 325(1): 41-49

Hwang C, Huang T (2006) Combustion synthesis of nanocrystalline ceria (CeO2) powders by a dry route. *Materials Science and Engineering:B* 132 (3) 229-238

Laosiripojana N, Assabumrungrat S (2005) Methane steam reforming over $Ni/Ce-ZrO_2$ catalyst: Influences of $Ce-ZrO_2$ support on reactivity, resistance toward carbon formation, and intrinsic reaction kinetics. *Applied Catalysis A: General* 290 (1-2): 200-211

Laosiripojana N, Assabumrungrat S (2006) Catalytic steam reforming of ethanol over high surface area CeO_2: The role of CeO_2 as an internal pre-reforming catalyst. *Applied Catalysis B: Environmental* 66 (1-2): 29-39

Laosiripojana N, Assabumrungrat S (2006) The effect of specific surface area on the activity of nano-scale ceria catalysts for methanol decomposition with and without steam at SOFC operating temperatures, *Chemical Engineering Science* 61 (8): 2540-2549

Lenka RK, Mahata T et al (2008) Combustion synthesis of gadolinia-doped ceria using glycine and urea fuels. *Journal of Alloys and Compounds* 466 (1-2): 326-329

Mahata T, Das G et al (2005) Combustion synthesis of gadolinia doped ceria powder. *Journal of Alloys and Compounds* 391(1-2) 129-135.

Nualpang W, Laosiripojana N et al (2008) Combustion synthesis of $Ce_{0.9}Gd_{0.1}O_{1.95}$ for use as an electrolyte for SOFCs. *Journal of Metal, Materials and Minerals* 18: 219-222

Patil KC, Aruna ST et al (2002) Combustion synthesis: an update, Current. *Opinion in Solid State and Materials Science* 6 (6): 507-512

Pino L, Vita A et al (2006) Performance of Pt/CeO_2 catalyst for propane oxidative steam reforming. *Applied Catalysis A: General* 306: 68-77

Ramirez-Cabrera E, Atkinson A et al (2002) Reactivity of ceria, Gd- and Nb-doped ceria to methane. *Applied Catalysis B: Environmental* 36 (3): 193-206

Sauvet AL, Fouletier J (2001) Catalytic properties of new anode materials for solid oxide fuel cells operated under methane at intermediary temperature. *Journal of Power Sources* 101 (2): 259-266

Skinner SJ, Kilner JA (2003) Oxygen ion conductors. *Materials Today* 6 (3): 30-37.

Xu H, Yan H et al (2008) Preparation and properties of Y3+ and Ca2+ co-doped ceria electrolyte materials for ITSOFC. *Solid State Sciences* **10**(9) 1179-1184.

Understandings of Solid Particle Impact and Bonding Behaviors in Warm Spray Deposition

M. Watanabe, K. H. Kim, M. Komatsu, S. Kuroda

National Institute for Materials Science, 1-2-1 Sengen, Tsukuba, Ibaraki, JAPAN 305-0047

1. Introduction

Coatings have become increasingly important to protect a substrate material from various degradation and damage such as wear, corrosion, and oxidation. Those protective functions lead to enhancement of durability of materials and to save total amounts of resources and energy consumptions. Thermal spraying is one of the most popular coating technologies because it can provide thick coatings over 100 μm over a large area in relatively much shorter deposition time compared with other coating processes such as electroplating, PVD and CVD. Among various thermal spray processes, Cold Spray (CS) and Warm Spray (WS) depositions are characterized by relatively high particle velocity (400 ~ 1000 m/s) and low particle temperature during deposition. In these processes, the temperature of sprayed particles can be lower than the melting point of the powder material.

Warm Spray (WS) deposition, which has been developed in National Institute for Materials Science (NIMS) in Japan, is new spray process. In conventional thermal sprayings such as high velocity oxy-fuel (HVOF) and plasma spraying (PS), feedstock particles are melted by combustion or plasma flame and deposited on a substrate. On the other hand, in WS, particles are not melted but moderately warmed up by controlling flame temperature, and then, impacted on a substrate with supersonic velocity to form a coating. Since the reduction of particle temperature can suppress detrimental reactions such as oxidation and decarburization during flight, it is possible for WS to fabricate a coating with high purity in air or a nanostructured coating by keeping the original microstructure of feedstock powder. Kawakita et. al. [1] investigated the oxide content of the titanium coatings deposited by WS. The oxide content varied depending on the process condition and the lowest value was comparable to the content in the feed stock powder. Watanabe et .al. [2] fabricated WC-Co coatings by WS deposition and compared with the microstructure and mechanical properties of the HVOF coatings. While the HVOF coatings contained the W_2C phases due to decarburization and the amorphous binder phases due to rapid melting in flame and quenching on the substrates, the WS coatings significantly suppressed the formation of those detrimental phases. The obtained coatings showed the improved fracture resistance at higher Co volume fraction due to higher ductility of remained Co binder phase.

However, the bonding mechanism of solid particle impact in CS and WS processes is not fully understood. It is generally accepted by various experimental results that there exists a critical impact velocity, at which a particle comes to bond with a substrate. Assadi et al. [3] performed numerical analyses to simulate solid particle impact and bonding behavior for cold spray process focusing on critical velocity. They calculated the stress and strain distribution, temperature increase during impact under adiabatic condition. The critical velocity was determined as the occurrence of numerical instability corresponding to adiabatic shear instability. Schmidt et. al. [4] further developed the numerical simulation of impact behavior and proposed the improved model to estimate critical velocity for various metals. The developed model showed good agreement with the experiments. Also Bae et. al. [5] carried out impact analysis of the different combination of material. However, since these studies were mainly targeted to simulate Cold Spray process, in which particle temperature is usually below 400 °C, the effects of particle temperature and thermal softening upon impact is not studied well. For Warm Spray process, it is essential to understand the effect of particle temperature and to reveal the correlation with the coating properties.

In the present research, numerical simulation of a titanium particle impact has been carried out by finite element method and the effects of particle temperature on the deformation behaviors and critical velocity were investigated. In addition, the effect of particle size was studied by applying thermal conduction model. Obtained results were directly compared with the cross sectional images of a single splat.

2. Analysis
2.1 Finite element model

In order to understand the effect of particle temperature on impact behavior, the finite element simulation was carried out by using a commercial finite element code (ABAQUS/explicit version 6.5). Figure 1(a) shows the axisymmetric model

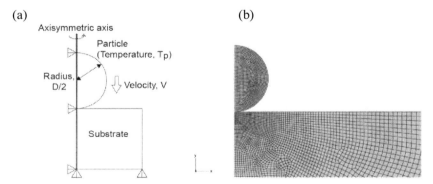

Fig. 1(a). Schematic of analysis model: a particle (diameter D, temperature T_p, velocity V) impacting on a substrate, (b) typical finite element mesh used in the simulation

Table 1. Parameters of titanium used in the analysis (for room temperature).

Parameter	unit	Ti
Density	kg/m^3	4507
Elastic modulus	GPa	110
Specific heat	J/kg·K	520
Melting Point	K	1941
Thermal expansion coefficient	10^{-6}/K	8.6
Yield stress, A	MPa	480
Strength coeffcient, B	MPa	1352
Strain rate coeffcient, C		0.04
Strain hardening component		0.27
Reference temperature	K	293
Reference strain rate	1/s	1
Thermal softening exponent		1

used in this study. A particle with diameter D is impacting on the substrate. The values of particle diameter was set to 30, 75, 125, 200 μm to investigate the effect of particle size. Three different particle temperatures $T_p = 293, 773, 1273$ K were applied for comparison. In Fig. 1(b), the typical mesh used in the analysis is shown for the initial mesh size of 0.5 μm near the contact surfaces. Four node elements with hourglass control were used. Johnson-Cook plastic deformation model was applied to express high strain rate plastic hardening of the materials [3-5]. In this model, strain hardening and thermal softening effects, and strain rate dependence can be taken into account.

$$\sigma = \left[A + B\varepsilon^n\right]\left[1 + C\ln\varepsilon^*\right]\left[1 - T^{*m}\right] \tag{1}$$

where σ is flow stress, ε is plastic strain, A is yield stress, B is strength coefficient, n is strain hardening exponent, C is strain rate coefficient, ε^* is normalized strain rate, and $T^* = (T - T_{ref})/(T_m - T_{ref})$ in which T_m is melting point, and T_{ref} is room temperature. Since most of coating formation process is based on the impact of a particle on the pre-deposited coating, the material properties of the particle and the substrate were assumed to be the same. The case of a titanium particle impacting on a titanium substrate was studied. The material properties are shown in Table 1. Note that most of parameters were set as a function of temperature and only the values at room temperature are shown in table 1. Apart from the typical simulation results reported so far [3-5], in the present study, the heat conduction was taken into account during whole simulation. According to Assadi et.al. [3], if a value of $x_{el}^2/D_{th}t$ is equal or above 1, the adiabatic assumption is justified based on the diffusive heat equation, where x_{el} represents an element size and D_{th} and t are the thermal diffusivity and the process time respectively. For the metals such as copper, aluminum, and titanium, the values of D_{th} is 117 x 10^{-6}, 96.8 x 10^{-6}, 9.25 x 10^{-6} m^2/s respectively. Substitution of typical values x_{el} = 0.5 μm and t = 10 ns, provides x_{el}^2/D_{th} = 0.22, 0.26, and 2.7 for those three metals. The values are smaller than 1 for copper and aluminum and larger than 1 for titanium. This estimation suggests that when very small microstructure in submicron scale is discussed, the effect of heat conduction during impact cannot be ignored. Although for titanium the value of x_{el}^2/D_{th} is larger than 1 and the adiabatic model might be reasonable enough, the heat conduction model was developed in the present study to construct more general numerical simulation procedure.

2.2 Critical velocity prediction

In the previous studies [6], it has been proposed that fracture of the surface oxide layers on a metal particle and a substrate and the formation of surfaces with high activation energy are necessary to bond each other. Based on this concept, in most of numerical simulation approaches, the occurrences of adiabatic shear instability which means sudden strain and temperature increase and flow stress break-up because of exceeding of the thermal softening rate over work hardening rate, or local melting due to large strain and high deformation rate, are typically applied as the bonding criterion. According to Assadi et.al., these two criteria provide similar result in adiabatic condition [3]. In the present study, the critical velocity is defined as the impact velocity at which the highest temperature in the particle can reach melting point of it. The temperature variations obtained from the numerical analysis are affected by the element size. Especially, when large deformation occurs in localized region, the effect of mesh size becomes significant. In general, the use of smaller element provides more reliable data but due to the limitation of computation power, the maximum temperature was estimated by linear extrapolation of the value for two element size models, 0.5 μm and 1.0 μm as elaborated later. The maximum temperatures were calculated for different impact velocity from 400 to 1200 m/s and the polynomial approximation curve was developed as a function of the impact velocity. From obtained formula and the melting point of titanium 1941 K, the critical velocity was finally decided. This procedure was carried out for different particle initial temperature and particle size.

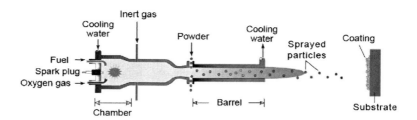

Fig. 2. Schematic of Warm Spray deposition.

Understandings of Solid Particle Impact and Bonding Behaviors

Table 2. Spray parameters

Parameter	unit	Ti Splat
Fuel flow rate	dm^3/min	0.3
Oxygen flow rate	dm^3/min	623
Nitrogen flow rate	dm^3/min	1500
Barrel length	mm	200
Spray distance	mm	180
Powder feed gas		Nitrogen
Substrate temperature	K	25
Spray gun traverse velocity	mm/	1500
Powder feed rate	g/min	4

3. Experiment
3.1 Warm Spray Deposition

Titanium particles were deposited on a titanium substrate at a low powder feeding rate of 4 g/min and a high spray gun transverse speed of 1500 mm/s via the warm spraying process so that particles did not overlapped each other. Commercially available titanium powder with the volume average of 28 μm (TILOP - 45μm, Sumitomo Titanium Corp., Tokyo, Japan) was used. Figure 2 shows the schematic of Warm Spray Process. In the process, a mixing chamber is equipped between a combustion chamber and a powder feeding port. The high temperature gas flow generated by combustion of kerosene and oxygen is mixed with nitrogen in order to lower gas temperature. By changing the flow rate of nitrogen, the temperature of gas flow can be controlled and thus the temperature of particles inputted from the feed port can be controlled and kept under their melting point. Then, accelerated and moderately heated-up solid particles impact on the substrate and can be bonded kinetically. In this study, nitrogen flow rate was 1.5 m^3/min, which is an optimized spraying condition for titanium particles [1,6]. The detailed spraying parameters were summarized in Table 2.

3.2 Microstructure observation of one splat

The cross-sectional thin sections of one titanium particle bonded on the substrate was prepared by utilizing focused ion beam (FIB) equipment (Hitachi FB-2100) and the prepared thin sample was observed by a Field Emission Transmission Electron Microscope (FE-TEM, JEOL JEM-2100F) with scanning TEM (STEM) mode operating at 200 kV. More detail explanation regarding sample preparation can be found in the references [5].

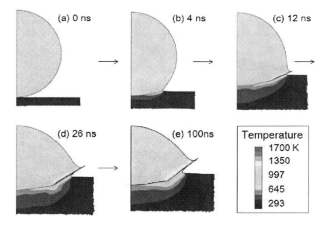

Fig. 3. Sequence of titanium particle deformation and temperature distribution during impact simulated for $T_p = 773$ K, $V_p = 800$ m/s, and $D = 30$ μm showing the formation of jet at the interface with temperature increase.

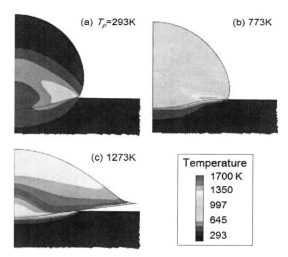

Fig. 4. Variation of particle deformation and temperature distribution at 200 ns after impact depending on the particle impact temperature T_p, (a) 293 K, (b) 773 K, (c) 1273 K for the particle velocity of 800 m/s and the diameter of 75 μm.

4. Results and Discussion

Figure 3 shows the deformation sequence of a titanium particle impacting on a titanium substrate calculated for the initial particle temperature of 773K, the impact velocity of 800 m/s, and the particle diameter of 30 μm. The formation of out-flowing jet can be recognized from the interface (Fig.3 (c)(d)(e)). The contour

Understandings of Solid Particle Impact and Bonding Behaviors 209

in the figures indicates the temperature distribution during impact. The highest temperature increase occurred near the edge of the contact region. In the Fig. 3(d), the highest value reached at about 1700 K at 26 ns after beginning of impact. The occurrence of huge shear flow in such short period resulted in large temperature increase of about 1000 K. On the other hand, there is little defamation and temperature increase at the epicenter region. Note that the area of high temperature region colored by red becomes smaller in Fig. 3(e) compared with Fig. 3(d) due to heat dissipation into the substrate and the particle.

Figure 4 shows the deformed particle shapes 200 ns later after beginning of impact for the different initial particle temperature 293, 773, 1273 K with the particle velocity of 800 m/s and the diameter of 75 μm. The contour indicates the variation of temperature. The aspect ratio, which is defined as the ratio of hight and diameter of a deformed splat, was 0.72, 0.59, 0.29 for T_p = 293, 773, 1273 K, respectively. Even though the impact velocity was the same value of 800 m/s for all three cases, the particle with T_p = 1273 K deformed 2.5 times more in terms of the aspect ratio. Since the temperature is the sum of the contributions of the initial temperature and the temperature increase due to heat generation during deformation and the reduction due to heat dissipation by thermal conduction, the highest temperature near the interface was achieved in the case of T_p = 1273 K. In the substrates, the heat affected area was the largest for T_p = 293 K because the particle was not thermally softened and the energy absorption by plastic deformation of itself was the lowest resulted in the largest deformation in a substrate. On the other hand, since this model takes into account the thermal conduction, the highest temperature in a substrate was also achieved for T_p = 1273 K.

In Figure 5, the effect of particle impact velocity on the maximum temperature during impact is plotted for D = 30 μm and T_p = 293 K as a function of the element size. As explained above, the estimated temperature is affected by the element size and the maximum temperature for each velocity was determined by extrapolating the line connecting the two data points obtained at 0.5 and 1 μm mesh size to zero. As reported in previous works [3], the maximum temperature increases for higher impact velocity. The extrapolated maximum temperature was replot-

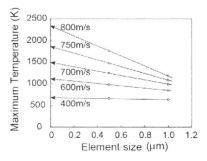

Fig. 5. Correlation of mesh size and the maximum temperature.

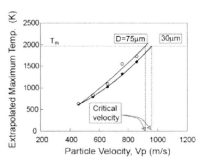

Fig. 6. Effect of particle size and velocity on the variation of the maximum temperature.

ted as a function of particle impact velocity for the particle size of 30 and 75 μm in Fig. 6. The initial particle temperature was 293 K. It can be seen that the larger particle size resulted in higher temperature with lower impact velocity and thus the critical velocity becomes lower for D =75 μm than 30 μm. This result suggests that if the impact velocity is the same, the impact of larger particle will increase temperature more around the interface and that the particle can be bonded with lower critical velocity. From the approximation curve in Fig. 6, the critical velocity was predicted as 914 m/s for 75 μm and 958 m/s for 30 μm. On the other hand, the difference is only 4.6 % for twice more particle size in the present titanium impact simulation. As discussed in the section 2.1, titanium has a low thermal diffusivity and the impact phenomena within submicron period might be treated as adiabatic condition. Under the adiabatic condition, the simulation results did not change for the different particle size due to lack of a factor which varied with surface area. Relatively small effect of particle size predicted in the present thermal conduction model for titanium case is not contradictory with the result of adiabatic model. In Fig. 7, the critical velocity of titanium particle impacting on titanium substrate is indicated as a function of particle temperature T_p and particle diameter D. Critical velocity rapidly decrease from 800 ~ 960 m /s for T_p = 293 K to 400 m/s for 1273 K. In Fig. 4, higher T_p leads to larger deformation and wider area of high temperature region at interface. Thus higher particle temperature can be expected to increase the deposition efficiency, to obtain larger bonded area and to decrease pores. Particle size showed relatively small effect compared with particle temperature. The difference in critical velocity is about 200 m/s between the smallest particle (30 μm) and the largest (200 μm) and the difference becomes small as the particle temperature increases and almost negligible over 773 K. As discussed above, this tendency would depend on the thermal diffusivity of a particle and a substrate. For a material of high thermal diffusivity such as copper and

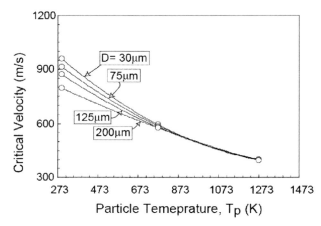

Fig. 7. Effect of particle size on critical velocity for titanium particle impacting on titanium substrate.

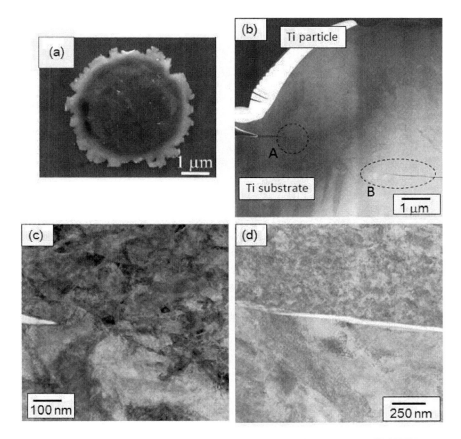

Fig. 8. (a) Example of a titanium particle deposited by Warm Spray process, (b) STEM cross sectional images of a titanium particle on a titanium substrate, (c) (d) higher magnification images of the region A and B in (b).

aluminum, particle size would have much larger effect on critical velocity and thus on microstructure and performance of their coatings.

Figure 8 shows the cross sectional images of actual titanium particle deposited on a titanium substrate by Warm Spray deposition. The TEM samples were made from the center regions of splats. The typical splat is shown in Fig 8a. The formation of jet-out region can be recognized at the edge region away from the epicenter as expected from the numerical simulation. In fact, the shape of the deposited titanium particle is similar to the analysis result in Fig. 3 implying adequacy of the simulation. In the region marked A, where the most severe shear deformation occurred, the particle and the substrate are intimately bonded (Fig. 8c). On the other hand, in the epicenter region marked B, where the shear instability did not occur, a gap between the particle and the substrate is recognized (Fig. 8d). According to the latest fluid dynamic simulation result of particle velocity and temperature in Warm Spray process, the velocity and the temperature were estimated as 800 m/s and 1039 K for the particle with 30 μm diameter. For the particle with $T_p \sim 1000$ K, the critical velocity can be estimated as 500 m/s in Fig. 7 assuming no effect of particle size. The particle velocity estimated by the simulation of Warm Spray process is well above the critical velocity and thus, the occurrence of the intimate bonding of a titanium particle for the spray condition in Warm Spray appears to be reasonable. In the present study, local melting of the interface is assumed as the criterion. There are several experimental reports implying the melting at the interface (for example, see ref. [7]), but in Fig.8 the clear evidence of the melting could not be recognized. Further investigation remains to reveal the criterion of bonding in Warm Spray deposition processes.

5. Conclusions

By applying finite element simulation with thermal conduction model, the impact behavior of a titanium particle deposited on a titanium substrate by Warm Spray deposition was simulated for various particle temperature, velocity, and size. The critical velocity was predicted as a function of those parameters and the result was compared with an actual splat observation result. The advantages of Warm Spray deposition, which can moderately heat up powder below its melting point, were quantitatively demonstrated.

References

[1]   Kawakita J, Kuroda S, Fukushima T, Katanoda H, Matsuo K, Fukanuma H.: Dense titanium coatings by modified HVOF spraying. Surf. Coat. Technol. 201, 1250-1255 (2006).
[2]   Watanabe M, Pornthep C, Kuroda S, Kawakita J, Kitamura J, Sato K.: Development of WC-Co Coatings by Warm Spray Deposition for Resource Savings of Tungsten, J. Jpn. Inst. Metals 71, 853-859 (2007).
[3]   Assadi H, Gärtner F, Stoltenhoff T, Kreye H.: Bonding mechanism in cold gas spraying. Acta Mater. 51, 4379-4394 (2003).
[4]   Schmidt T, Gärtner F, Assadi H, Kreye H.: Development of a generalized parameter window for cold spray deposition. Acta Mater. 54, 729-742 (2006).
[5]   Bae G, Xiong Y, Kumar S, Kang K, Lee C.: General aspects of interface bonding in kinetic sprayed coatings. Acta Mater. 56, 4858-4868 (2008).
[6]   Kim KH, Watanabe M, Kuroda S.: Bonding mechanisms of thermally softened metallic powder particles and substrates impacted at high velocity. Surf. Coat. Technol. 204, 2175-2180 (2010).
[7]   Bae G, Kumar S, Yoon S, Kang K, Na H, Kim H-J, Lee C.: Bonding features and associated mechanisms in kinetic sprayed titanium coatings. Acta Mater. 57, 5654-5666 (2009).

Mechanical properties of innovative metal/ceramic composites based on freeze-cast ceramic preforms

Siddhartha Roy, Jens Gibmeier, Kay André Weidenmann, Alexander Wanner

Institut für Angewandte Materialien, Karlsruher Institut für Technologie, Kaiserstr. 12, 76131 Karlsruhe, Germany

Abstract

This article provides an overview of recent research carried out about the mechanical properties of an innovative metal/ceramic composite fabricated by squeeze-casting liquid Al-12Si in freeze-cast alumina preforms. The composite consists of domains made of alternating metallic and ceramic lamellae. Elastic and elastic-plastic deformation behavior and mechanism of internal load transfer under external compressive stresses in individual domains with different orientations were studied. Results show that the stiffness is highest along the freezing direction and lowest along the direction perpendicular to it. When compressed along directions parallel to freezing direction, the composite is strong and brittle. When compressed along other directions, the behavior is controlled by the soft metallic alloy. Studies of internal load transfer under external compressive stress along different directions show that the mechanism of internal load transfer differs significantly when loading direction changes from that along the freezing direction to the direction perpendicular to it.

1 Introduction

Motivations behind the usage of metal matrix composites (MMC) are threefold [1]: i) MMCs allow going beyond the boundaries drawn in property space by monolithic metals and ceramics, ii) making a MMC is the only way of introducing significant amount of an oxide or a carbide into an important metal and iii) incorporation of finely divided ceramics in metallic matrices allows to utilise the advantageous properties of the ceramic materials. MMCs can be classified according to the form of the reinforcement (such as fibres, particles, whiskers or short fibres, interpenetrating etc). While fibre reinforced composites are anisotropic, particle or short fibre reinforced and interpenetrating MMCs are mostly isotropic [2]. An excellent review of the different MMC fabrication routes can be found in Ref. [3].

Further advancement towards MMCs with novel architecture has recently been made in the form of metal/ceramic composites based on freeze-cast ceramic preforms. Here, an open porous ceramic preform is first made by freeze-casting of a ceramic slurry, freeze-drying and subsequent sintering [4, 5]. These open porous ceramic preforms have excellent permeability and sufficient strength for melt infiltration and they can be infiltrated with liquid metal by squeeze-casting or even die-casting [6]. Because of the anisotropic growth kinetics of the ice crystals, the preforms fabricated from water based slurries have a hierarchial lamellar domain structure [5]. After melt infiltration, the structure consists of domains having alternating ceramic and metallic lamellae. The lamellae spacing as well as the size of individual domains are dependent upon the parameters of freeze-casting [6].

A brief overview of recent research carried out on metal/ceramic composites fabricated by squeeze-casting Al-12Si melt in freeze-cast alumina preforms is given in this work. The main aim of the research was to investigate the mechanical behavior of the individual domains having different orientations. To reach this goal, single-domain samples were first prepared from poly-domain samples. Elastic properties of the single-domains were studied using ultrasound phase spectroscopy while elastic-plastic flow behavior as well as the internal load transfer mechanism by energy dispersive synchrotron X-ray diffraction was studied under external compression.

2 Experimental procedure

2.1 Specimen material

Porous alumina preforms having approximately 56 vol % porosity were fabricated at Institut für Keramik im Maschinenbau (IKM) at the Karlsruhe Institute of Technology, Germany by freeze-casting of slurry containing alumina powder, freeze-drying and subsequent sintering. The slurry had 22 vol % ceramic powder and it was freeze-cast at -10° C. Detail description of the preform fabrication is given in Ref. [7]. Preforms with nominal dimensions 10×44×66 mm³ were infiltrated with Al-12Si melt by squeeze-casting at the Materials Technology Section of Aalen University of Applied Sciences, Germany. The left hand side image of Fig. 1 shows the optical micrograph of such a poly-domain sample for the face perpendicular to the freezing direction. In this image, the darker regions correspond to the ceramic phase while the brighter regions belong to the metallic alloy. Individual domains with alternating ceramic and metallic lamellae can be observed and they have also been marked in the image. Single-domain samples were prepared from poly-domain samples by diamond wire cutting and metallographic polishing of the poly-domain samples. A diamond coated steel wire having 220 μm diameter was used for this purpose. Microstructures of two such single-domain samples for the face perpendicular to the freezing direction are shown in Fig. 1 (right hand side). The single-domain samples had individual edge lengths in the range of 1.8-2.6 mm.

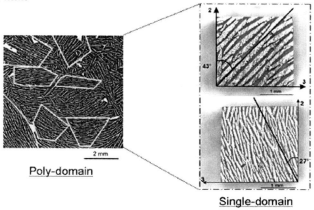

Fig. 1: Optical micrograph showing a poly-domain (left) sample and two single-domain (right) samples. The micrographs show the faces perpendicular to the freezing direction of the preform.

2.2 Elastic analysis

Elastic analysis of the single-domains was carried out by ultrasound phase spectroscopy (UPS). Thorough theoretical description of the technique can be found in Ref. [8], while detail description of application of this technique for the material under study is given in Ref. [9]. In this technique the sample is held between two identical broadband transducers (longitudinal or shear for the respective elastic constants). Continuous, harmonic, sinusoidal waves are passed through the sample and its phase shift is measured as a function of frequency. From the slope of the phase vs. frequency spectrum the velocity of the propagating sound wave can be measured, which can then be used along with sample density to determine the elastic constant.

2.3 Compression test and study of internal load transfer mechanism

Compression test of the single-domains were carried out using a miniature mechanical test machine from Kammrath & Weiss GmbH, Dortmund, Germany. Two hardened steel punches were used for load application and 20 µm thick aluminum foils were used between the sample and the punches to minimise frictional problems. Compression tests were carried out at a fixed crosshead velocity of 2 µm/s, corresponding to a nominal strain rate of 10^{-3}/s.

Study of internal load transfer mechanism under external compression was carried out using energy dispersive synchrotron X-ray diffraction. The experiments were carried out at materials science beamline EDDI at BESSY, Berlin. Thorough description of the experimental procedure is given in Ref. [10]. The same compression test setup described above was used for this purpose. This test setup was mounted on the sample positioning table of the diffractometer unit. The diffraction angle was kept constant at $2\theta = 7°$. The gauge volume within the sample was defined by the dimensions of the slits on the path of the incoming and outgoing beams. Compressive stress on the sample was increased stepwise, and at each step diffraction measurements were carried out following the $\sin^2\psi$ technique by tilting the loading rig and thus the sample in 20 steps between $\psi = 0\text{-}90°$. Three single-domain samples having approximately $0°$ orientation were studied. One of these three samples was compressed along the freezing direction, the second one was compressed along $0°$ orientation, while the last sample was compressed along $90°$ to domain orientation. In all cases a preload in the range of 20-30 N (corresponding to a stress of 3-5 MPa) was first applied to make sure that the sample did not fall down during subsequent tilting, and the lattice spacings corresponding to this initial preloaded state were used as reference values for strain calculation.

3 Results and discussion

Fig. 2 shows the effect of domain orientation on the longitudinal elastic constants of single-domain samples measured transverse to the freezing direction. Detailed analysis of the elastic behavior of the composite has been studied in Ref. [9]. In this image the open circles correspond to the experimentally determined values. Although there is a large scatter, the plot clearly shows that the domain orientation has a strong influence on the stiffness of individual domains. The stiffness decreases from approximately 220 GPa at orientations close to $0°$ to approximately 165 GPa at orientations close to $90°$. The lines in the plot correspond to predictions from the Postma model for 3D laminates with alternating layers of random thickness [11] for different values of E-modulus for alumina. Input parameters for the model are the E-moduli of the metallic and the ceramic phases and their respective volume fractions. Ceramic content in each sample was determined from mass-dimension analysis and assuming absence of porosity. The average ceramic content for all samples were 42 vol%. Because of their random poly-crystalline nature, both alumina and Al-12Si have been assumed to be isotropic. E-modulus of Al-12Si was determined during the current study and it was 80 GPa. E-modulus of alumina was unknown beforehand. Hence, a value of 390 GPa was taken from literature. Corresponding predictions are shown by the solid line in Fig. 2.

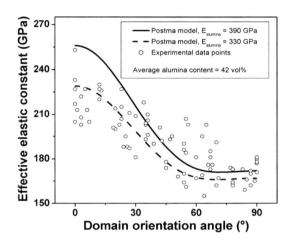

Fig. 2: Effect of domain orientation on the stiffness of single-domain samples measured along directions transverse to the freezing direction.

The distribution in Fig. 2 clearly shows that at orientations < 30°, the theoretical predictions overestimate the experimental predictions. To get a better match between theory and experiment, E-modulus of alumina was varied, keeping other parameters fixed. It is observed that at $E_{alumina}$ = 330 GPa, the match between theoretical predictions and experimental observations are much better. This may probably be explained by the fact that due to the presence of defects within the ceramic preform, its stiffness is significantly less from its literature value.

Fig. 3 shows the compressive stress-strain plots for single-domain samples compressed along the freezing direction, along approximately 4° to domain orientation and along approximately 87° to domain orientation. Compressive stress-strain plot for unreinforced Al-12Si is also shown as a reference. When compressed along the freezing direction, the behavior is dominated by the ceramic phase. The sample shows high compressive strength and limited or no plasticity. The initial curved regions in each plot at very low stresses are artefacts attributed to the sample friction and non-parallelism of the contact faces. The single domain sample with approximately 4° domain orientation also shows similar ceramic controlled brittle behavior. However, the sample with approximately 87° domain orientation shows metallic alloy controlled behavior. Because of the presence of the ceramic lamellae, its strength is significantly higher than the unreinforced Al-12Si. It has been reported in Ref. [12] that domain orientation has a significant effect on the compressive behavior of individual domains. At domain orientations in the range of 0-30° the composite behavior is ceramic controlled, while at orientations between 30-90° the behavior is metallic alloy controlled.

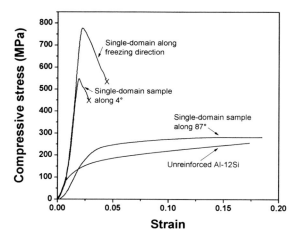

Fig. 3: Compressive stress-strain plot for single domain samples. Stress-strain plot for unreinforced Al-12Si is shown as a reference.

Measured lattice spacings along and transverse to the loading direction in all three phases (Al soild solution, Si and alumina) were used to determine the longitudinal and transverse microstrain at various applied stresses. As, the lattice spacings corresponding to the initial applied preload were used as the reference, hence the measured microstrains were in fact changes with respect to this reference state. Hence, here onwards microstrains correspond to "change in microstrain" within each phase. Fig. 4 shows the change in lattice microstrains along the loading direction in aluminum solid solution and alumina in the three single domain samples. In these plots the continuum mechanics average lattice microstrains in each phase are shown. These were calculated from measured lattice microstrains in several diffracting planes of each phase and using their respective multiplicity factors (refer to Ref. [10] for details). The error bars in each plot correspond to the standard deviations among lattice microstrains from various diffracting planes within a phase and they directly correlate to interplanar anisotropy.

Fig. 4 shows that the microstrain evolution in alumina and aluminum is similar in the samples compressed along freezing direction and along 0°. In both cases the microstrains in aluminum initially increase until about -100 MPa applied stress, and at higher stresses, it reaches an almost constant value. This happens due to plastic deformation of aluminum and subsequent load transfer to alumina at higher stresses. As alumina carries most of the load, microstrain within it continuously increases with increasing applied stress. In the sample compressed along 90° orientation, microstrain in alumina initially increases similar to the other two samples. However, at applied stresses higher than about -150 MPa, it becomes almost constant, while lattice microstrain within aluminum increases. Following Fig. 3, this change of behavior corresponds to the transition from macroscopic elastic to plastic deformation.

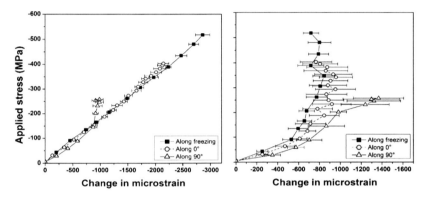

Fig. 4: Evolution of change in microstrain in alumina (left) and aluminum (right) as a function of applied compressive stress in the three single domain samples

Previous study on progressive damage evolution [12] has already shown that when loaded along orientations between 45-90°, transverse cracks start to develop within alumina at this transition region due to tensile mismatch between the metallic and the ceramic phases. As alumina starts to crack, its load bearing capacity decreases, and hence more load is carried by aluminum.

4. Conclusions

A brief overview of the mechanical behavior of single domains of a metal/ceramic composite fabricated by squeeze-casting Al-12Si in freeze-cast alumina preforms is given in this work. Following conclusions can be drawn:

- Stiffness of the domains are highest along the freezing direction and they are lowest along directions perpendicular to it.
- Mechanical behavior under compression changes from ceramic controlled strong and brittle along the freezing direction to metallic alloy controlled soft and ductile along directions orthogonal to it.
- Load transfer mechanism is strongly dependent upon the direction of load application with respect to the freezing direction.

Acknowledgements

We thank T. Waschkies, R. Oberacker, and M.J. Hoffmann of the Institute for Ceramics in Mechanical Engineering, University of Karlsruhe, for the processing of the ceramic perform as well as A. Nagel and co-workers at Aalen University of Applied Sciences, Aalen, Germany, for final processing of the metal ceramic composites by squeeze casting. Financial support from the Deutsche Forschungsgemeinschaft (DFG) under projects Wa1122/3-1 and Wa1122/4-1 is gratefully acknowledged.

List of References

[1] Mortensen A, Llorca J (2010) Metal matrix composites. Annu Rev. Mater. Res. 40: 243-270
[2] Chawla N, Chawla KK (2006) Metal matrix composites. Springer, New York
[3] Lloyd DJ (1997) in Mallick PK (ed) Composites Engineering Handbook, Marcel Dekker Inc.
[4] Deville S, Saiz E et al (2006) Freezing as a path to build complex composites. Sci. 311: 515 – 518
[5] Deville S, Saiz E et al (2007) Ice-templated porous alumina structures. Acta Mater. 55: 1965 – 1974
[6] Mattern A (2005) Metall-Keramik-Verbundwerkstoffe mit Isotropen und Anisotropen Al_2O_3 Verstärkungen. Dissertation: Universität Karlsruhe (TH)
[7] Waschkies T, Oberacker R et al (2009) Control of lamellae spacing during freeze-casting of ceramics using double-sided cooling as a novel processing route. J Am Ceram Soc. 92: S79 – S84
[8] Wanner A (1998) Elastic moduli measurements of extremely porous ceramic materials by ultrasonic phase spectroscopy. Mater. Sci. Engg. A248: 35 – 43
[9] Roy S, Wanner A (2008) Metal/ceramic composites from freeze-cast preforms: domain structure and mechanical properties. Comp. Sci. Technol. 68: 1136 – 1143
[10] Roy S, Gibmeier J et al (2009) In-situ study of internal load transfer in a novel metal/ceramic composite exhibiting lamellar microstructure using energy dispersive synchrotron X-ray diffraction. Adv. Engg. Mater. 11: 471 – 477
[11] Postma GW (1955) Wave propagation in a stratified medium. Geophysics. 20: 780 – 806
[12] Roy S, Butz B et al (2010) Damage evolution and domain-level anisotropy in metal/ceramic composites exhibiting lamellar microstructures. Acta Mater. 58: 2300-2312

Mini-samples technique in tensile and fracture toughness tests of nano-structured materials

Tomasz Brynk*, Rafal M. Molak, Zbigniew Pakiela, Krzysztof J. Kurzydlowski

Warsaw University of Technology, Faculty of Materials Science and Engineering

Woloska 141, 02-507 Warsaw, POLAND

*e-mail: tbrynk@inmat.pw.edu.pl

Abstract Samples dimensions defined in standards for tensile and fracture toughness tests may be too large in the case of modern materials produced in small volumes, e.g. nano-structured metals. Also, dimensions of irradiation tests samples are frequently not appropriate for standardized samples. In such situations so called mini-samples need to be used. The paper presents methodology and the results of tensile and fracture toughness tests which were carried out on mini-samples with a few millimeter dimensions. These samples were made of nano-structured metals processed by hydro extrusion (HE) and equal channel angular pressing (ECAP). Due to small size of specimens optical method of strain measurement - digital image correlation (DIC) - was applied. DIC two-point-tracing mode was used as an optical extensometer in the tensile tests. The inverse method was applied for precise determining stress intensity factors and crack tip positions from displacement fields acquired by DIC. Comparison of the results of the tensile and fracture toughness tests which were carried out on standardized and mini-samples made of the same materials is also presented and the results are discussed in terms of the applicability of mini-samples technique in the studies of nano-metals.

Introduction

Tensile tests are one of the most commonly used methods of characterizing mechanical properties of materials. It is relatively simple, economic and fully standardized. It provides such widely employed parameters as elastic limit, ultimate strength, fracture stress, elongation at fracture. The slope of linear part of elastic range of stress-strain plot can also be used to estimate Young modulus. The result obtained from tensile tests are routinely used for selection of materials and designing the dimensions of parts for a given application.

Materials and various components may have cracks and discontinuities, being the consequence of production method and/or in service degradation, which should be considered from the point of view of damage tolerance. In the context, the parameter describing stress fields near the elliptical crack tip is stress intensity factor. If some requirements concerning samples dimension related to crack length are fulfilled, critical stress intensity factor is obtained which is the material characteristic. There are standards defining geometry of samples with artificial cracks and testing methods for determining critical stress intensity factor. If the loading is cyclic there is a possibility to determine crack propagation rate as a function of the applied stress intensity factor and so called Paris' plot.

In many research fields of modern engineering there are situations in which standard geometry samples are not possible to machine. An example of such a situation is testing mechanical properties of small parts or welds. Another example are new nano-structured metals obtained by severe plastic deformation (SPD) which are currently in the size of millimetres. Also coupons for irradiation test are frequently too small for cutting standardized samples. Determination of the mechanical properties of such small samples/parts calls for new methods, among others use of sa called mini-samples.

The aim of the project described here was to validate the use o mini-samples in tensile, fracture toughness and crack growth rate tests. To this end results obtained for standard samples have been compared with the ones obtained on mini-sample. Also, mini-samples have been employed to characterize mechanical properties of SPD nano-metals.

One of the challenges in testing with mini-samples is an accurate strain measurement. There are no available small electromechanical extensometers for samples with gauge sections under 5mm. One way of solving this difficulty is to perform optical strain measurements via Digital Image Correlation (DIC) [4,10,11]. The method was used here for strain measurements due to the ease in sample preparation, the accuracy offered by megapixels cameras and insensitivity to vibrations made by servo hydraulic testing stands (contrary to laser interferometry methods).

Another difficulty related to mini-samples is a scale effect. When materials, devices and structures are scaled to reduce their size many of their properties, including the mechanical, thermal, optical, electrical and magnetic change with respect to their macroscopic equivalents. Dependence of the properties on the size of the item is defined as the "scale effect." Literature sources indicate the occurrence of at least a few types of mechanisms of the scale effect which are related with different physical phenomena and can have various effects on the mechanical properties of elements under consideration. From an engineering point of view, taking into account the mechanisms responsible for the different action of the scale effect, one can make the following division [5]:

- micro structural effects;
- statistical effect of grain;
- the effect of strain gradient;
- the effect of boundaries and surfaces;
- geometric effect.

Another classification, which can be found in the literature, is into: (a) inner and (b) outer scale effect [1]. Internal scale effects result from microstructure features, such as the size of grains and second phase particles and distance between the particles. In principle, to fully assure dimensional proportions, reduction in the size of the test samples would require reduction in the size of the grains, particles and distances between them.

In the case of a static tensile test of polycrystalline metals the important parameter affecting the strength is the thickness of the sample and the number of grains per thickness.

The reaction of a small sample to the applied stress will also depend on the "dimensionality" of the applied stress. In the case of deformation of samples in the presence of so-called strain gradient (torsion, bending) scale effect is very clearly visible.

The value of the parameters describing strength of the materials generally increase with the decreasing size of the sample [6,13]. A similar effect has been observed in the case of hardness measurements [8].

In conclusion, there are a number of challenges in scaling down size of the test samples and their use requires experimental verification, which has been a subject of the present communication.

Digital Image Correlation (DIC)

Optical methods of strain measurements have become popular in the recent years because of the development of image acquisition systems and computer hardware with higher performance. The main advantage of these methods is possibility of tracing the position of a large number of test points on the sample surface unlike electromechanical extensometers which provide information on the relative displacements of "two-points". Optical methods can therefore be used to obtain displacement fields map and provide much more information on the macro- and micro-scale mechanics of plastic deformation.

Digital Image Correlations [4,10,11] is one of the most popular optical strain measurement techniques. The set-up consists of a digital camera which registers images in gray levels scale, sources of white light and PC computer images storage and software for analyzes.

The data analyses are based on comparison of the digital images of sample surface before and after deformation. Speckle patterns produced by spraying white and black paint is required on the observed surface for appropriate contrast. Small rectangle subsets of the area of interest have unique distribution of gray levels and their positions may be traced using correlation function. Application of interpolation functions improves displacement measurement accuracy which may be up to 0.01pixel.

Commercially available software Vic2d [14] was used in this study for DIC measurements. Also, a 2D DIC two point tracing (optical extensometer) was used. Displacement maps were used for calculations of the stress intensity factors.

Tensile tests

Samples and method

Mini-samples has been designed based on the standard [12], however, with the scaling factor of 100. Shape and dimensions of the samples are shown in Figure 1.

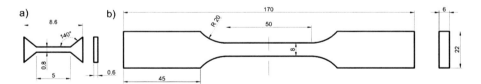

Fig.1. Samples for tensile tests: a) mini, b) standard (scale 1:4)

The samples were cut by electric discharge machining (EDM) from a rod of Al 5483 series alloy with a diameter 50 mm. Static tensile tests were carried out using two testing machines: static MTS QTest/10 and dynamic MTS 810. All samples were strained at a constant strain rate $1 * 10^{-3}$ [1/sec]. For the analysis of macroscopic strain DIC optical extensometer was used.

Results

Table 1 shows the mean values of the mechanical parameters obtained for standard and mini-samples.

Table 1 Mechanical properties of standard and mini-samples

Samples type	Parameter	$R_{0,2}$ [MPa]	R_m [MPa]	A^c [%]	n	A_{gt} [%]	ΔA_n [%]	R_u [MPa]
Standard	Mean	227	386	10.2	0.243	8.65	2.03	372
	Std. deviation	8.9	5.4	1.1	0.01	1.6	0.77	3.9
Mini	Mean	245	406	12.1	0.229	9.22	3.06	363
	Std. deviation	4.8	7	1.1	0.0055	0.66	1.2	11

Figure 2 shows the representative stress-strain curves for both types of samples.

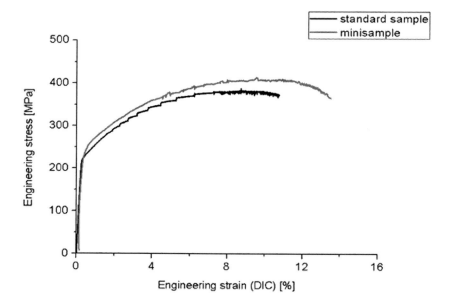

Fig.2. Stress-strain curves for standard and mini-samples

It can be noted from the curves in Figure 2 that for mini-samples an increase in tensile and yield strength has been observed. Strain to failure (A^c) and strain localization (ΔA_n) is also higher for mini- than for standard sample. The results are contrary to the so-called surface model [7].

Analyzing the data in Figure 2 one should take into account that with the decreasing size of samples there is an increase in their surface-to-volume ratio. Thus, if surface is more resistant to deformation, for example because of oxidation, an increase in strength with the decreasing size of the samples is expected. Such a surface induced increase was reported for compression of single crystal beams [**Fehler! Verweisquelle konnte nicht gefunden werden.**]. In the present study, surface hardening may also be related to the defects generated during the machining of mini-samples (EDM cutting). This, however, requires further study.

Fracture toughness test

Method for determining SIF from DIC results

Stress intensity factor (SIF) describes stress and strain or displacement fields near the elliptical crack tip. If plain strain conditions are satisfied the critical value of SIF, just before unstable crack growth, is the material constant.

Measurements of SIF are performed on samples with artificially machined notches. The samples are cyclically loaded to obtain pre-crack with sharp edge. Then they are monotonically loaded to fracture registering the applied force and displacement on their edges. The measured displacements are used to calculate actual crack length. SIF is calculated using equations of fracture mechanics.

SIF may be also determined from displacement fields near the crack tip by so called inverse method. The results presented here were obtained using the procedure based on Newton-Raphson method [15] for fitting the parameters of the following equations:

$$u_x = \sum_{n=1}^{\infty} \frac{K_{In}}{2G\sqrt{2\pi}} r^{n/2} \left(\kappa \cos\frac{n}{2}\theta - \frac{n}{2}\cos(\frac{n}{2}-2)\theta + (\frac{n}{2}+(-1)^n)\cos\frac{n}{2}\theta \right)$$
$$- \sum_{n=1}^{\infty} \frac{K_{IIn}}{2G\sqrt{2\pi}} r^{n/2} \left(\kappa \sin\frac{n}{2}\theta - \frac{n}{2}\sin(\frac{n}{2}-2)\theta + (\frac{n}{2}-(-1)^n)\sin\frac{n}{2}\theta \right)$$

$$u_y = \sum_{n=1}^{\infty} \frac{K_{In}}{2G\sqrt{2\pi}} r^{n/2} \left(\kappa \sin\frac{n}{2}\theta - \frac{n}{2}\sin(\frac{n}{2}-2)\theta - (\frac{n}{2}+(-1)^n)\sin\frac{n}{2}\theta \right)$$
$$- \sum_{n=1}^{\infty} \frac{K_{IIn}}{2G\sqrt{2\pi}} r^{n/2} \left(-\kappa \cos\frac{n}{2}\theta - \frac{n}{2}\cos(\frac{n}{2}-2)\theta + (\frac{n}{2}-(-1)^n)\cos\frac{n}{2}\theta \right)$$

where:

$$\kappa = \frac{3-v}{1+v} \text{ for plane stress, or } \kappa = 3-4v \text{ for plane strain,}$$

$r = \sqrt{(x-x_0)^2 + (y-y_0)^2}$ and (x_0, y_0) are coordinates of the crack tip,

G –bulk modulus, v- Poisson's ratio.

Fitting procedure was applied in the Matlab code. As the input data it required initial values of K_I, K_{II}, x_0, y_0 and to read previously exported data of u_x and u_y from DIC measurements, a new iterative procedures was used which reduces the estimation error.

For assuring the same geometry of the area of measurements while the crack develops, the subarea of interest was shifted during the tests. This subarea was set to at 50% of height and 50% of width of the total DIC measurements area and the crack tip position was set at half of its height and half of the width. As a result, the fitting process was performed always on the same size area and the crack tip was kept in the same position with regard to the frame for all analyzed images. Also data points laying near the crack line were removed from calculation due to error related with speckle pattern damage in this region.

Test procedure

Comparative test were carried out on mini-samples made of Al 5483 alloy shown in Figure 3a and on compact specimens in Figure 3b. Standardized compact samples were used not only for measurements with clip-on-gage extensometer (compliance method). However, in the same tests DIC measurements were performed.

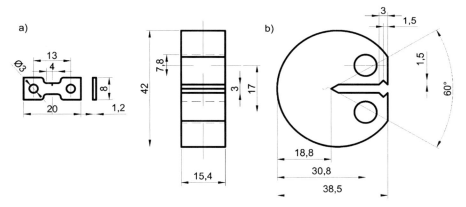

Fig.3. Samples for fracture tests: a) mini , b) compact

Mini-samples were tested on a horizontal testing stand coupled with camera and optical lens for DIC measurements. Before the tests pre-cracks were generated. After calibrating cameras orientation, capturing of images started synchronously with the loading. Images were registered at 0.5 s time intervals. Data from optical measurements were exported to ASCII files and processed by Matlab, as described earlier.

Compact samples were tested according to standard for fracture toughness measurements [2]. During the test images of previously painted sample were recorded and processed by the same algorithm as for mini-samples.

Results

Experiments proved for standard samples plain stress conditions. Therefore it was not possible to determine fracture toughness. The measured K values are plotted vs. normalized crack length (a/W) for the tested samples in Figure 4. The examples of fitting are presented in Figure 5 and 6. Displacement maps with fitted boundaries are presented on white background which shows the total area of total DIC measurements.

Optical measurements curves for mini- and standardized samples have similar shape. The translation in a/W axis is related with higher initial values for normalized crack length in the case of mini-samples. Differences between optical and compliance methods for standard samples are related to the plastic zone at the crack tip, which has influences crack length measurements and crack closure at the beginning of each test.

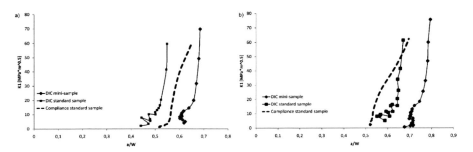

Fig. 4. Comparison of the obtained results for mini- and standard samples with crack oriented a) perpendicular to the rod direction, b) parallel to the rod direction

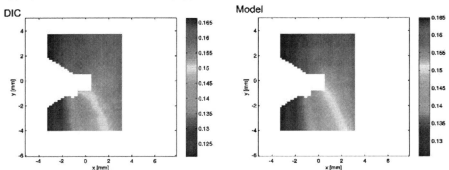

Fig. 5. Results of the fitting process for selected moments of the test for standard sample (vertical displacement)

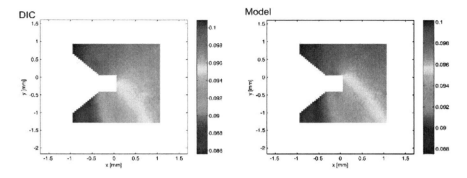

Fig. 6. Results of the fitting process for selected moments of the test for mini-sample (vertical displacement)

Threshold SIF tests

Test procedure

DIC based crack growth rate investigations on mini-samples made of Al 5483 alloy after ECAP process were reported in [3]. Present paper presents the results of the threshold SIF (K_{th}) measurements. Sample used to this end had the same dimensions as the ones used for crack resistance curves. The same methodology was applied for SIF and crack length measurements.

The first step of the test procedure was to generate the initial crack of 2 mm length by cyclical loading with maximal force F_{max}=410 N, R=0.1 and f=20Hz . After that sample was unloaded and initial image of its surface was taken. Then the loading was decreased to F_{max}=150 N with unchanged other parameters. Such conditions were maintained for 20 000 cycles. After that, the maximum force was increased by 10N and cyclic loading carried out for additional 20 000 cycles. Such procedure was continued up to the fracture of the sample. Every 1000 cycles, test was stopped on at the maximum force to register the image for DIC measurements.

Results

The result of the test are shown in Figure 7. These results were obtained for measurement of SIF and crack position from images registered every 1000 cycles. Crack length reached minimum for K=4.9MPa*(m)$^{1/2}$ and after achieving this value started to increase. K value at the minimum crack length was interpreted as K_{th}.

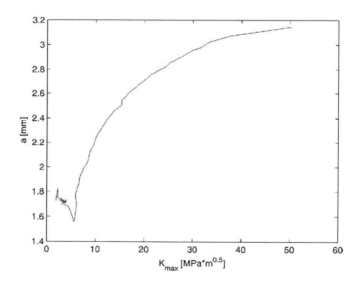

Fig.7. Results of K_{th} measurements

Conclusions

Mini-samples tests give the opportunity to measure mechanical properties of small parts and materials which are not available in the volume sufficient for preparing standard ones.

The results presented here show applicability of sub-sized samples to tensile, fracture toughness and crack growth rate tests with optical displacement measurements. The precision and reproducibility of the data allow for verification of scale effect models.

Tensile tests gave very similar results for standard and mini-samples with the scale effect having some influence on the obtained results. Optical measurements of SIF allowed to accurate determine the K curves in plane stress conditions for mini-samples, which had similar shape but are translated to higher values of normalized crack length comparing to standard ones. Good resolution of optical measurements allowed for crack closure effect registration and determination of threshold SIF.

Acknowledgments: This work was supported by the Polish Ministry of Science and Higher Education under the contract no. R15 024 03.

References

1. 1. Arzt E (1998) Size effects in materials due to microstructural and dimensional constraints: a comparative review, Acta Materialia 46:5611-5626
2. ASTM E399 standard
3. Brynk T, Rasinski M, Pakiela Z, Olejnik L, Kurzydlowski KJ (2010) Investigation of fatique crack growth rate of Al 5484 ultrafine grained alloy after ECAP process, Phys Status Solidi A 207:1132–1135
4. Chu TC, Ranson WF, Sutton MA, Peters WH (1985) Applications of Digital Image Correlation Techniques to Experimental Mechanics, Exp Mech 25:232–244
5. Janssen PJM, de Keijser ThH, Geers MGD (2006) An experimental assessment of grain size effects in the uniaxial straining of thin Al sheet with a few grains across the thickness, Mater Sci and Eng A 419:238–248
6. Fleck NA, Muller GM, Ashby MF, Hutchinson JW (1994) Strain gradient plasticity: theory and experiment, Acta Metallica Materialia 42:475-487
7. Engel U, Eckstein R (2002) Microforming—from basic research to its realization, Journal of Mater Process Technol 125-126:35-44
8. Ma Q, Clarke DR (1995) Size dependence of the hardness of silver single crystals, Journal of Mater Res 10:853-863
9. Norfleet DM, Dimiduk DM, Polasik SJ, Uchic MD, Mills MJ (2008) Dislocation structures and their relationship to strength in deformed nickel microcrystals, Acta Materialia 56:2988–3001
10. Peters WH, Ranson WF (1982) Digital imaging techniques on experimental stress analysis, Opt Eng 21:427-431
11. Peters WH, Ranson WF, Sutton MA, Chu TC, Anderson J (1993) Application of digital image correlation methods to rigid body mechanics, Opt Eng 22:738–742
12. PN-EN 10002-1 standard

13. Stölken JS, Evans AG (1998) A microbend test method for measuring the plasticity length scale, Acta Materialia 46: 5109-5115.
14. www.limess.de
15. Yoneyama S, Morimoto Y, Takashi M (2006) Automatic Evaluation of Mixed-mode Stress Intensity Factors Utilizing Digital Image Correlation, Strain 42:21–29

The use of Focused Ion Beam to Build Nanodevices with Graphitic Structures

B. S. Archanjo[1], E. H. Martins Ferreira[1], I. O. Maciel[2], C. M. Almeida[1], V. Carozo[1], C. Legnani[1,4], W. G. Quirino[1,4], C. A. Achete[1,3] and A. Jorio[1,4]

[1] Divisão de Metrologia de Materiais, Instituto Nacional de Metrologia, Normalização e Qualidade Industrial (INMETRO), Duque de Caxias, RJ, 25250-020, Brazil

[2] Departamento de Física, ICE, Universidade federal de Juiz de Fora, Juiz de Fora, MG, 36036-330, Brazil

[3] Programa de Engenharia Metalúrgica e de Materiais (PEMM), Universidade Federal do Rio de Janeiro, Cx. Postal 68505, CEP 21945-970, Rio de Janeiro, RJ, Brazil

[4] Departamento de Física, Universidade Federal de Minas Gerais, Belo Horizonte, MG, 30123-970, Brazil

Abstract

The modification of samples using focused ion beam (FIB) is a very powerful technique in many areas of material science, especially on modification and construction of nanodevices. The aim of this work is the creation of defects, fabrication of ordered patterns and direct deposition of Pt contacts on graphitic structures (from few layers graphene to many layers graphite) by using a Ga^+ FIB source together with a field emission gun scanning electron microscope (FEG-SEM) in a dual beam platform. Using this platform, FIB capabilities for fabrication of nanodevices for scientific and technological development are investigated. Micro-Raman Spectroscopy was used to track the changes caused by these fabrication processes by analyzing the ratio between the defect induced Raman D band and the structural G band. This approach provides information about the performance and the damages caused by dual beam techniques when used on graphene samples for device applications.

1 Introduction

Dual beam platforms are powerful techniques to change material properties at nano- and microscale. These equipments have a focused ion beam (FIB) and a scanning electron microscope (SEM) gun and they have also commonly available a gas injection system to perform chemical vapor deposition [1-6]. Both electron and ion columns have high resolution performance and well controlled spot positioning. These features allow controlled visualization, deposition and removal of materials. This technique then improves our knowledge about controlled modifications of physical systems at nano- and microscale ranges and also about the use of nanopatterning and nanostructuring to fabricate nanodevices [1-6]. In special, FIB and SEM are used as tools to create amorphous defects on graphitic structures.

Different methods are also applied to understand the interactions of ions and electrons with graphitic structures, including atomic force microscopy (AFM), micro-Raman spectroscopy (RS) and computer simulation [7-13].

Raman spectroscopy (RS) is one of the most used techniques to probe the presence of defects and even quantify the amount of disorder created by ion implantation on graphitic materials [11-21]. RS is a very sensitive, fast and non-destructive way to get information about the different phases of a material. For instance, it can be used to undoubtedly distinguish a monolayer from a bilayer graphene, or a semi-conducting from a metallic nanotube. Since RS gets information about the vibrational modes of a material, it can also be used to probe the deformation of the crystal lattice caused by defects, such as vacancies, adatoms or dopants. In the Raman spectra of any graphitic material there is always a peak around 1500 cm^{-1}, called G band, which is related to the in-plane stretching mode of the C-C bonds. This is the only first order Raman scattering process allowed in a perfect (defect free) system. However, when there is a symmetry breaking in the system due to a presence of a defect such as a vacancy or a border, one can also observe two new features around 1360 cm^{-1} and 1620 cm^{-1}, which are called D and D' bands respectively. The ratio between the intensities of the D and G bands (I_D/I_G) can be used to monitor the density of defects in the material. For instance, in HOPG, I_D/I_G is inversely proportional to the micro crystallite size [18-21]. As the amount of disorder in the crystal structure increases, the G and D bands get broader until the observed spectra becomes a single broad band characterizing an amorphous phase.

This work reports improvements in understanding the performance of Ga$^+$ ion surface implantation using a FIB source at small level of modification. For instance, the use of a small ion current and a fast dwell time (fast writing speed) were studied to modify graphitic structures with low damage for nanodevices applications. A silicon substrate and SEM images were used to obtain the fastest dwell time in which the FIB is capable to define a correct pattern. Using this suitable dwell time, a periodic pattern was then defined in a graphene sheet. AFM and RS were used to understand and to measure the amount of defects created. Electrical contacts were also made via electron beam induced chemical vapor deposition on a graphene sheet where the goal is to verify the change in graphene structure due to the damage made by the ions.

2 Experiment

The graphite-based structures were prepared by a single mechanical exfoliation of the HOPG, purchased from NT-MDT company (HOPG ZYB grade, 20x20x2 mm) specimen and deposited onto a Si substrate with a 300 nm layer of SiO$_2$ [22]. The samples were found optically and the number of layers identified by both Raman spectroscopy [23] and AFM. The bombardments and FEG-SEM high-resolution images were performed using a Nova Nanolab dual beam platform from FEI, where SEM images were acquired by using a 10 keV accelerating energy and 0.13 nA electron current. The ion bombardments were executed using a Ga$^+$ ion source working at 30 keV accelerating energy, with an ion current of about 1.6 pA and different dwell times (writing speeds), which can go down to 100 ns per dot. According to the Dual Beam manufacturer, the smaller FWHM beam diameter using this current is about 7 nm. AFM analysis was carried out with a Nano Wizard AFM (JPK) operated in intermittent contact mode, where topographic and lock-in phase images were recorded simultaneously. Micro-Raman scattering measurements were performed by using a Horiba Jobin-Yvon T64000 triple-monochromator spectrometer in the backscattering configuration, equipped with N_2 cooled CCD detector and a microscope with a 100x objective. The excitation laser energy was 2.41 eV (514.5 nm). The micro-Raman images were collected using a Witec Alpha300 AR Atomic Force and Confocal Raman Microscopy System equipped with a 532 nm exciting laser allowing confocal Raman imaging with a lateral resolution limited by the beam diameter to ~360 nm using a 100x objective.

3 Results and Discussion

As discussed before, graphitic structures have some characteristics that offer exciting promises for future electronic devices. Nevertheless, making high quality electronic devices from graphene, which typically involves etching it or material deposition into nanoscale structures, has proved to be a challenging task [24]. In Figure 1 we show some schematic examples of periodic patterned modifications, which leads to further unexpected and potentially useful charge transport behavior [24].

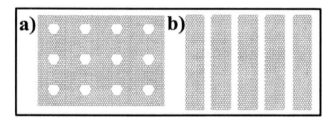

Figure 1: Schematic diagram of graphene superlattice. a) Two-dimensional graphene superlattice of periodic spots. b) One-dimensional graphene superlattice formed by periodic lines.

With the development of FIB technology, such structures can be obtained in an easy and fast way mainly for basic science research, providing further experimental progress for testing different device configurations, which may point the directions that the microfabrication industries will follow.

3.1 Defects on graphitic structures and performance of focused Ion beam

The amount of defects caused by a 30 keV Ga^+ incident beam on graphitic structures was analyzed by the evolution of the Raman disorder parameter I_D/I_G, which is the ratio of the intensities between the D band and G band, as a function of ion bombardment dose for graphite with one, two, three layers, 5 nm thick graphite and HOPG, as shown in Figure 2.

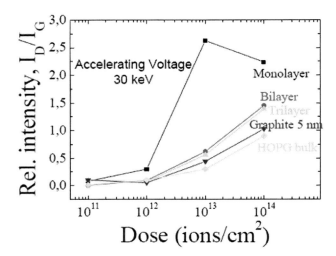

Figure 2: Intensity ratio (I_D/I_G) taken from Raman spectra plotted as a function of ion bombardment dose for mono-, bi-, trilayer graphene, 5 nm thick graphite and bulk HOPG.

The maximum I_D/I_G values observed in bilayer and trilayer graphene are more similar to the behavior of bulk graphite structure than to the behavior of graphene single layer, despite their similarity to the latter structure. This result shows how the bombardment has different influence on the sample structure as we change from one monolayer to a few layer graphene. The evolution of disorder caused by the 30 keV Ga^+ ions on single-layer graphene is similar to those caused by low energy ions [22]. As the ion dose increases, there is an increase, saturation and finally a decrease of the I_D/I_G ratio for the single-layer graphene. Such behavior is not observed for the multilayer graphene or HOPG, whose I_D/I_G ratios only increase with increasing ion dose and get smaller when the number of layers increases.

To visualize the performance of the ion beam at low fluence doses, we performed four squared spot arrays where the shortest dwell time during patterning is 100 ns. This means that an image collected at this rate has the highest scan frequency provided by the equipment. For a pattern produced using this dwell time at the lowest available beam current (~ 1.6 pA) just one ion per dot is delivered at each spot, in average. However, with a 100 ns dwell time, the time that the beam takes to go from one spot to another is of the same order of magnitude of this dwell time, causing the ions to be spread out along the beam path. Figure 3 shows these four patterns made on a silicon dioxide substrate where the dwell time and the number of passes were varied in order to keep the number of ions per dot constant at about 10^6 ions per dot. In Figure 3a the dwell time is 4.6 ms and the pattern is well defined on the substrate. As the dwell time is reduced down to 1 μs the spots become distorted, but still can be observed (Figure 3b). For a dwell time of 0.5 μs (Figure 3c) a path between the spots can be seen. Finally, for a dwell time of 100 ns, the original pattern is not exactly the intended one and what we see is a continuous line which is the beam path. In this case, the array of spots is lost. However, the lines defined by the beam path are well defined.

Analyzing these four patterns it can be noticed that for an array of spots a dwell time higher than 1 μs must be used, once this dwell time provides a satisfactory accuracy. However if the intended pattern is an array of lines, even the dwell time of about 100 ns can be used.

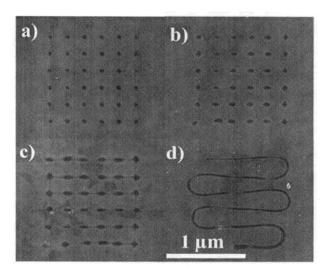

Figure 3: High resolution FEG-SEM images of a silicon surface bombarded by a Ga+ FIB working at 30 keV and a current of about 7 pA. Each spot is created by ~ 106 ions per dot using different dwell times and number of passes on silicon substrate. (a) 5 passes and 4.6 ms dwell time. (b) 23000 passes and 1 μs dwell time (c) 46000 passes and 0.5 μs dwell time. (d) 230000 passes and 0.1 μs dwell time.

3.2 Patterning graphene with FIB

The focused ion beam was used to create a pattern with parallel lines on graphene single layer aiming to obtain a periodic structure. During the bombardment with Ga^+ ions to create the lines, ion implantation and backscattering occurs in the sample, resulting in undesired defects. The structure was analyzed after the bombardment by AFM and RS in order to verify the amount of defects created. During the patterning, a current of ~1.6 pA was employed in Ga^+ ions delivering about 10 ions per each 10nm of the patterned line. The beam diameter is expected to be 10 nm. The patterned lines are spaced from each other by 200 nm and they can be seen in Figure 4. An optical image of the graphene sheet is shown in Figure 4a. The blue square drawn in the image shows the patterned region. In Figure 4b it can be seen an AFM phase image showing the bombarded region and the topography image of this same region is shown in Figure 4c. In the AFM images the lines can be easily seen and it can be noted that the pattern regions are very well defined, suggesting that just a small amount of ions fall outside the patterned lines, creating just few random defects.

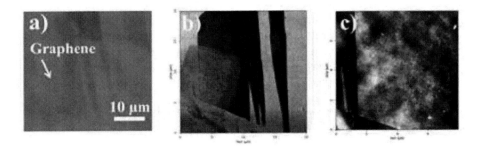

Figure 4: (a) Optical image of the graphene sheet deposited onto SiO2/Si substrate. The blue square defines the FIB pattern region. AFM phase (b) and topography (c) images after ion lithography.

The RS maps of the integrated intensity of the G band and the D band can be seen in Figure 5a and 5b, respectively. The D band map shows that the defects were created only in the desired area, demonstrating that FIB can pattern a specific spot of a device without influencing the other parts of the structure. In other to study the orientation of the defects in the patterned lines, we change the polarization of the incident beam during micro Raman experiments [25]. The Raman spectra of the graphene sheet for the laser polarization parallel and perpendicular to the patterned lines can be seen in Figure 5c. One can observe that the intensity ratios I_D/I_G are different for each polarization. This corroborates that the patterned lines are well defined because one would expect that a perfect armchair border would show a D band only when the incident laser polarization is parallel to it [25]. Therefore, as the results show that the lines created by the ion beam have a smaller D band when the laser polarization is perpendicular to the lines than when the laser polarization is parallel, the lines are not completely diffuse.

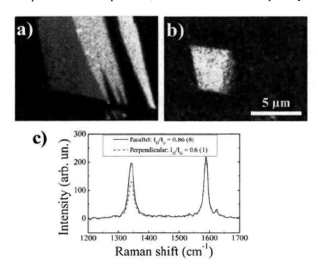

Figure 5: Raman maps plotting the integrated intensity of the graphene (a) G band and (b) D band while scanning the sample. (c) Raman spectra of the defected region using the laser polarization parallel and perpendicular to the patterned lines.

3.3 Electrical contacts on graphene

Pt electrical contacts were made by using electron beam induced chemical vapor deposition on graphene. This makes it possible to study the conductivity of the system. The deposition was performed using a 2 keV electron accelerating energy in order to reduce the amount of defects during this procedure. Studies where Pt electrical contact is deposited on graphitic structures can be found in the literature [26, 27], since Pt deposition has a better metallic behavior [28]. However, ion deposition creates more defects when compared to electron deposition, mainly because the ion collision at 30 keV on graphitic structure is more effective on removing carbon atoms [10,14]. Figure 6a shows an optical image of these contacts where the contact in one side of the sample was build on the top of the graphene layer and the other contact is placed on graphite piece. A zoom of the same region can be seen in the SEM image shown in Figure 6b. In both images no defects or changes in graphene structure can be recognized out of the area where Pt was deposited. A Raman spectrum of the sample was made before and after the contact

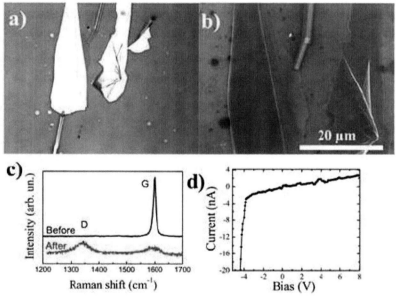

Figure 6: (a) Optical image and (b) SEM image of graphene with electrical contacts of Pt deposited by e-beam chemical vapor deposition. (c) Micro-Raman spectra of this graphene layer before and after Pt deposition. (d) Electrical characterization of the Pt contacted graphene layer.

deposition, as shown in Figure 6c. As RS is very sensitive to probe defects in graphitic structures, defects were easily detected in the graphene structure after the bombardment, since a very intense and broad D band is observed. This indicates a high level of disorder, but as the G peak is still significant the graphene structure was not destroyed. This can be explained by the fact that the interaction of the injected gas and electron beam at the graphene surface causes the deposition of contaminants on the whole region of the sample, increasing the intensity of D peak, meaning that the measured defects may not come from changes on the structure of the graphene itself. Electrical current measurements were also performed and the I x V curve is shown in Figure 6d. The Pt contacts deposited on the graphene sample presented a Schottky like behavior, which shows that the electron beam chemical vapor deposition method is not the optimum technique to define contacts on graphene.

4 Conclusions

This work reports improvements on understanding controlled ion bombardment, nanopatterning and micro fabrication on HOPG, mono, bi and trilayer graphene by making a series of tests on FIB and by studding the created defects on the samples, aiming the creation of future technological standards for industry and for carbon-based material characterization. The samples were also analyzed by scanning probe microscopy, scanning electron microscopy and Raman spectroscopy. RS has shown to be more sensitive to study the created modifications, showing defects that no other technique could detect. The I_D/I_G ratio has a different evolution behavior for graphene and few layer graphene as the ion dose increases. Pt contacts were deposited on graphene using chemical vapor deposition induced by electron beam. Its electrical properties were investigated, showing a Schottky-like behavior, which is not expected for good quality devices, making it necessary to look for another deposition methods, such as conventional electron-beam lithography.

Acknowledgments

The authors acknowledge the financial support from the Brazilian Agencies CNPq, FINEP and FAPERJ.

References

[1] Seliger, R. L., Kubena, R. L, Olney, R. D., Ward, J. W., Wang, V.: High-resolution, ion-beam processes for microstructure fabrication. J. Vac. Sci. Technol. **16**, 1610-1612 (1979).

[2] Giannuzzi, L. A.: Introduction to focused ion beam – Instrumentation, theory, techniques and pratice Springer, New York (2005).

[3] Tseng, A. A.: Recent developments in micromilling using focused ion beam technology. J. Micromech. Microeng. **14**, R15–R34 (2004).

[4] Reyntjens, S., Puers, R.: A review of focused ion beam applications in microsystem technology. J. Micromech. Microeng. **11**, 287–300 (2001).

[5] Stokes, D. J., Wilhelmi, O., Reyntjens, S., Jiao, C., Roussel, L.: New methods for the study and fabrication of nano-structured materials using FIB SEM. J. Nanosci. Nanotechnol. **9**, 1268-1271 (2009).

[6] Hernandez-Ramırez, F.; Rodriguez, J.; Casals, O.; Russinyol, E.; Vila, A.; Romano-Rodrıguez, A.; Morante, J. R.; and Abid, M. Sensors and Actuators B 2006, 118, 198.

[7] Gierak, J.: Focused ion beam technology and ultimate applications. Semicond. Sci. Technol. **24**, 043001 (2009).

[8] O'Donnell, S. E., Buettner, M., Reinke, P.: Characterization of focused ion beam induced defect structures in graphite for the future guided self-assembly of molecules. J. Vac. Sci. Technol. B **27**, 2209-2216 (2009).

[9] Melinon, P., Hannour, A., Bardotti, L., Prevel, B., Gierak, J., Bourhis, E., Faini, G., Canut, B.: Ion beam nanopatterning in graphite: characterization of single extended defects. Nanotechnology 19, 235305 (2008).

[10] Teweldebrhan, D., Balandina, A. A.: Modification of graphene properties due to electron-beam irradiation. Appl. Phys. Lett. **94**, 013101 (2009)

[11] Krasheninnikov, A. V., Nordlund, K.: Ion and electron irradiation-induced effects in nanostructured materials. J. Appl. Phys. **107**, 071301 (2010).

[12] Dresselhaus, M. S., Kalish, R.: Ion implantation in diamond, graphite and related materials. Berlin, Springer-Verlag (1992) and references contained therein.

[13] Lopez, J. J.; Greer, F.; Greer, J. R.: Enhanced resistance of single-layer graphene to ion bombardment. J. Appl. Phys. **107**, 104326 (2010).

[14] Zhou, Y. B.; Liao, Z. M.; Wang, Y. F.; Duesberg, G. S.; Xu, J.; Fu Q.; Wu, X. S.; Yu, D. P.: Ion irradiation induced structural and electrical transition in graphene. The J. Chem. Phys. **133**, 234703 (2010).

[15] Tuinstra, F., Koenig, J. L.: Raman Spectrum of Graphite. J. Chem. Phys. **53**, 1126-1130 (1970).

[16] Ferrari, A. C., Robertson. J.: Interpretation of Raman spectra of disordered and amorphous carbon. Phys. Rev. B **61**, 14095 (2000).

[17] Jorio, A., Lucchese, M. M., Stavale, F., Achete, C. A.: Raman spectroscopy study of Ar+ bombardment in highly oriented pyrolytic graphite. Phys. Status Solidi B **246**, 2689-2692 (2009).

[18] Krasheninnikov, A .V., Nordlund, K., Keinonen, J.: Energetics, structure, and long-range interaction of vacancy-type defects in carbon nanotubes: Atomistic simulations. Phys. Rev. B: Condens. Matter **65**, 165423 (2002).

[19] Nakamura, K., Kitajima, M.: Ion-irradiation effects on the phonon correlation length of graphite studies by raman-spectroscopy. Phys. Rev. B: Condens. Matter Mater. Phys. **45**, 78-82 (1992).

[20] Cançado, L. G., Takai, K., Enoki, T., Endo, M., Kim, Y. A., Mizusaki, H., Jorio, A., Coelho, L. N., Magalhaes-Paniago, R., Pimenta, M. A.: General equation for the determination of the crystallite size L-a of nanographite. Appl. Phys. Lett. **88**, 163106 (2006).

[21] Lucchese, M. M., Stavale, F., Martins Ferreira, E. H., Vilani, C., Moutinho, M. V. O.,Capaz, R. B., Achete, C. A., Jorio, A.: Quantifying ion-induced defects and Raman relaxation length in graphene by Raman spectroscopy. Carbon 48, 1592-1597 (2010).

[22] Geim, A. K., Novoselov, K. S.: The rise of graphene. Nat. Mat. **6**, 183-191 (2007).

[23] Ferrari, A. C., Meyer, J. C., Scardaci, V., Casiraghi, C., Lazzeri, M., Mauri, F., Piscanec, S., Jiang, D., Novoselov, K. S., Roth, S., Geim, A. K.: Raman Spectrum of Graphene and Graphene Layers. Phys. Rev. Lett. 97, 187401 (2006).

[24] Park, C. H., Yang, L., Son, Y. W., Cohen, M. L., Louie, S. G.: Anisotropic behaviours of massless Dirac fermions in graphene under periodic potentials Nat. Phys. **4**, 213-217 (2008)

[25] Cancado, L. G, Pimenta, M. A., Neves, B. R. A., Dantas. M. S. S. Jorio, A.: Influence of the atomic structure on the Raman spectra of graphite. Phys. Rev. Lett. **93**, 247401 (2004)

[26] Shailos, A.; Nativel, W.; Kasumov, A.; Collet, C.; Ferrier, M.; Gueron, S.; Deblock, R.; Bouchiat, H.: Proximity effect and multiple Andreev reflections in few-layer graphene. Europhys. Lett. **79**, 57008 (2007).

[27] Shao, Q.; Liu, G.; Teweldebrhan, D.; Balandina, A. A.: High-temperature quenching of electrical resistance in graphene interconnects. Appl. Phys. Lett. **92**, 202108 (2008).

[28] Fernández-Pacheco, A.; Teresa, J. M.; Córdoba, R.; Ibarra, M. R.: Metal-insulator transition in Pt-C nanowires grown by focused-ion-beam-induced deposition. Phys. Rev. B **79**, 174204 (2009).

Development of compact continuous-wave terahertz (THz) sources by photoconductive mixing

H. Tanoto[1], J. H. Teng[1], Q. Y. Wu[1], M. Sun[2], Z. N. Chen[2], S. J. Chua[1,3], A. Gokarna[4], J. F. Lampin[4], and E. Dogheche[4]

[1]*Agency for Science, Technology and Research (A*STAR), Institute of Materials Research and Engineering, 3 Research Link, Singapore 117602, Singapore*

[2]*Agency for Science, Technology and Research (A*STAR), Institute for Infocomm Research, 1 Fusionopolis Way, #21-01 Connexis (South Tower), Singapore 138632, Singapore*

[3]*National University of Singapore, Centre for Optoelectronics, E3 #02-07, Engineering Drive 3, Singapore 117576, Singapore*

[4]*Institut d'Electronique de Microélectronique et de Nanotechnologie, CNRS, France*

Abstract

The electromagnetic terahertz (THz) frequency range is typically defined as 0.1 THz to 10 THz. This frequency range remains as the least developed part of the EM spectrum range, due to the lack of efficient emitters, detectors and transmission technology. Nonetheless, myriad applications of THz waves have been found or proposed, for example: spectroscopy, ultra large bandwidth communications, astrophysics and atmospheric science, biological and medical imaging, security screening and illicit material detection, and non-destructive evaluation. A highly desirable component for many applications is a compact, coherent, continuous-wave (cw) solid-state source of THz radiation. Semiconductor-based devices have been known to be highly compact and efficient, as apparent in the productions of integrated-circuits (ICs) and diode lasers. Unfortunately, approaches using traditional semiconductors electronics devices are limited by the significant roll off in output power at such high frequencies due to both transit-time and resistance–capacitance (RC) effects. On the other hand, photonic approaches are limited by the lack of semiconductor materials with sufficiently small bandgaps (1 THz ~ 4 meV). Recently, more techniques to generate THz waves using semiconductor materials are developed. One of such techniques is the photoconductive mixing (photomixing) of two near infrared lasers with THz beat frequency in ultra-fast semiconductor materials. This technique is promising for realization of compact, semiconductor-based and efficient CW THz emitters. This paper discusses on our efforts to develop compact and efficient continuous-wave THz photomixers.

Introduction

Efficient and portable emitters and detectors of electromagnetic (EM) waves with terahertz (THz) frequency are highly desirable for real-life systems in astronomy, biomedical, and homeland security [1]. Semiconductors have been historically employed in highly compact and efficient EM sources and detectors. Nonetheless, conventional semiconductor device technologies are limited to generating EM waves outside the desired THz frequency. This is either due to the RC time delay in the electronic devices or due to the lack of meV bandgap in the interband semiconductor optoelectronic materials. It is apparent that a different approach is needed to generate THz waves using semiconductor materials. One of the most promising approaches is the photoconductive mixing or photomixing [2]. In this technique, two optical EM waves from two lasers are superimposed onto ultra-fast semiconductor materials. The optical pump induces changes in conductivity in the materials and generates current upon introduction of externally biased voltage to the material. The current components are divided into DC and AC. The AC frequency matches the difference between optical frequencies of the two lasers. Setting this frequency difference in THz range means the AC current oscillate in THz too. Therefore it can be used to drive a planar metal antenna to radiate the continuous-wave (CW) THz waves into space.

In order to achieve a highly efficient operation, the materials and planar antenna designs have to be optimized. The materials used in photomixing must exhibit high resistance, high carrier mobility and ultra-short carrier lifetime [3]. Low-temperature grown GaAs (LT GaAs) has been shown to exhibit such characteristics [4] and therefore is our materials of choice. We employed molecular beam epitaxy (MBE) to grow LT GaAs under different growth and annealing conditions to obtain desired materials characteristics. Subsequently, CW THz photomixers were fabricated, characterized and implemented into an imaging system.

Experimental Details

Employing MBE, 1 μm of LT GaAs and 1 μm of AlAs heat spreading layer were deposited on a 2-inch semi-insulating GaAs substrate. The nominal growth temperature for LT GaAs is 200°C. Hall Effect measurements revealed a resistivity of 5x107 Ohm.cm and room temperature carrier mobility of 5000 cm^2/V.s. Employing a pump-probe technique with a 100-fs 812nm Ti-sapphire laser, the carrier lifetime of the LT GaAs was found to be 880fs.

A CW THz photomixer device was fabricated using a standard photolithography and electron beam lithography (EBL) processes. The planar antenna structure used in this work is a dual-dipole THz antenna [5] consisting of Ti/Au metals. Acting as a current source to drive the antenna is an optical injection region with interdigitated electrodes. There are eight electrode fingers, each with 200nm width, and with 800nm gap between adjacent fingers. The entire optical injection area is approximately 5μm x 8μm. These sub-micron electrodes are to ensure efficient photocarrier capture. Figure 1 shows a scanning electron microscope (SEM)

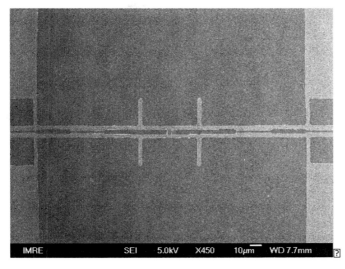

Fig. 1: Plan view scanning electron microscope (SEM) image of a THz photomixer with dual-dipole antenna

plan view image of the fabricated device. The device was then mounted on a hyper-hemispherical lens made of high-resistivity silicon. Keithley sourcemeter was employed to supply the voltage bias and monitor the photocurrent in the circuit. The fiber end from two tunable DFB lasers operating at around 850nm was aligned at the optimum position above the optical injection region of the photomixer. THz emission was detected using a fourier transform infrared spectroscopy (FTIR) system in conjunction with a liquid helium cooled silicon bolometer.

Results and Discussions

As the difference frequency of the two lasers was tuned from 0.2 – 1.8 THz, the device emits THz was corresponding to the difference frequency. The THz emission was detected as single peak by the FTIR. From this data, both the intensity of the THz emission as well as the peak THz frequency is known. However, to find the output power of the device, we had to calibrate the bolometer using a known blackbody source. In this work, a commercial Hg arc lamp supplied with the FTIR system was used. Between 0.2-2THz, the Hg lamp was found to act as an ideal Blackbody radiation with 4000K temperature. We calculated the radiated power of the lamp by performing integration of the Planck radiation formula over the desired frequency range. Using the Hg arc lamp power information, the output power of the device could be found by dividing the device integrated bolometer intensity at specific frequency range with those of the Hg arc lamp.

Employing this method, we found that the device displays THz emission tunable from 0 – 1.4 THz with peak emission frequency at 0.4THz with output power of 250nW. The THz output power at 1THz was approximated to be 10nW. We further investigate the characteristics of the device by varying the bias voltage and the optical input power. The following relationship was obtained from the equivalent circuit model of a THz photomixer [6]:

$$P_{THz}(\omega) = \frac{2R_A V_B^2 m P_1 P_2}{A} \left[\frac{\eta \mu e}{rh\nu} \right]^2 \frac{\tau^2}{1+(\omega\tau)^2} \frac{1}{1+(\omega R_A C)^2} \quad (1)$$

where P_{THz} is the THz emitter output power, R_A is the antenna resistance, V_B is the bias voltage, m is the modulation constant, P_1 and P_2 are the optical pump power, A is the active area, η is the external quantum efficiency, μ is the carrier mobility, e is the electron charge, r is the width of antenna fingers, $h\nu$ is the photon energy, τ is the carrier lifetime, ω is the difference frequency, and C is the capacitance of the photomixer. From (1), the THz output power should have a proportional quadratic relationship with both optical input power and the biased voltage. Figure 2a

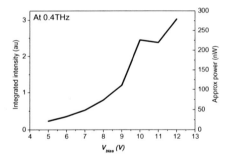

Fig: 2a: Output power of the photomixer at 0.4THz as function of applied bias

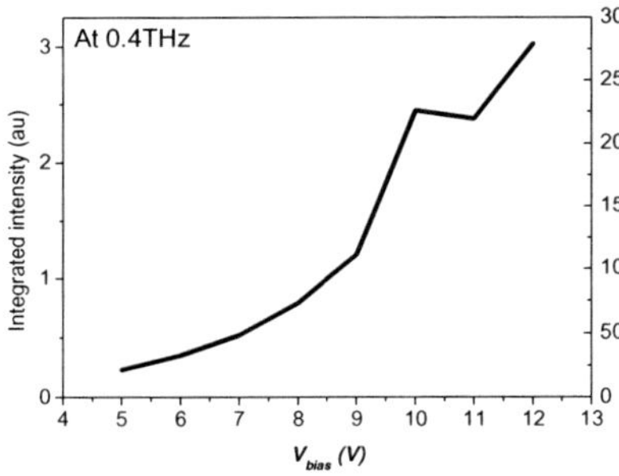

Fig: 2b: Output power of the photomixer at 0.74THz as function of optical input power

and 2b show the change in the device output power as function of bias voltage and optical input power, respectively. From the trends in both figures, it clearly in accordance with the equivalent circuit model. In addition, the THz output power saturation was observed at optical input power beyond 90mW. At approximately 85mW of optical input power, the efficiency of the photomixer reached its maximum, as indicated by the highest point in Figure 2b. With input power exceeding 85mW, the number of carrier generated does not increased but instead the heat generated in the materials started to affect the efficiency of the photomixer.

We finally implemented the device as a source in a preliminary CW THz imaging experiment. The device was tuned to emit at 0.4THz, at which it exhibits the

Development of compact continuous-wave terahertz (THz) sources 249

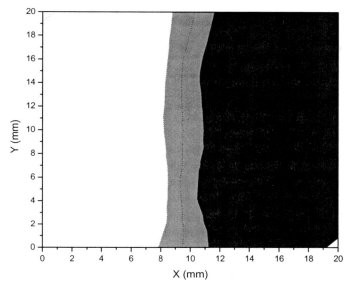

Fig. 3: Preliminary THz image of a metal plate obtained by mounting the plate on XY stage in front of the photomixer emitting at 0.4THz

highest output power. An XY stage was placed in between the source and the FTIR window. A metal plate was mounted on this stage and was raster scan to obtain the THz image. The resulting THz image is presented in Figure 3. Each point represents a 5mm step of the stage and the intensity information from each point was plotted into a contour plot. This experiment demonstrates the feasibility of using the device as light source for THz imaging.

Conclusions

We have demonstrated a CW THz photomixer employing LT GaAs as the material and dual-dipole planar antenna to radiate the THz waves. The device performances were discussed with emphasis on the output power response with changing bias voltage and optical input power. Finally, the device was implemented into an imaging system. Further improvements on the output power are expected through implementations of novel planar antenna designs and new active region designs with high carrier capture efficiency.

Acknowledgements

This work is funded by the Agency for Science, Technology and Research (A*STAR), Singapore, under grant number 0821410038.

References

[1] B. Ferguson and X. C. Zhang, "Materials for terahertz science and technology," Nature Materials, 1, pp. 26-33, (2002).

[2] E. R. Brown, K. A. McIntosh, K. B. Nichols, and C. L. Dennis, "Photomixing up to 3.8 THz in low-temperature-grown GaAs", Appl. Phys. Lett. 66, 285 (1995).

[3] I. S. Gregory, C. Baker, W. R. Tribe, M. J. Evans, H. E. Beere, E. H. Linfield, A. G. Davies, and M. Missous, "High resistivity annealed low-temperature GaAs with 100 fs lifetimes," App. Phys. Lett., 83, pp. 4199-4201, (2003).

[4] K. A. McIntosh, E. R. Brown, K. B. Nichols, O. B. McMahon, W. F. DiNatale, and T.M. Lyszczarz, "Terahertz photomixing with diode lasers in low-temperature-grown GaAs," App. Phys. Lett., 67, pp. 3844-3846, (1995).

[5] S. M. Duffy, S. Verghese, K. A. McIntosh, A. Jackson, A. C. Gossard, and S. Matsuura, "Accurate modeling of dual dipole and slot elements used with photomixers for coherent terahertz output power," IEEE Trans. Microwave Theory Tech., 49, pp. 1032-1038 (2001).

[6] I. S. Gregory, C. Baker, W. R. Tribe, I. V. Bradley, M. J. Evans, E. H. Linfield, A. G. Davies, and M. Missous, "Optimization of Photomixers and Antenna for Continuous-Wave Terahertz Emission", IEEE Journal of Quantum Electronics, 41, 717 (2005).

Electrical Impedance Characterization of Cement-Based Materials

Supaporn Wansom

National Metal and Materials Technology Center

114 Thailand Science Park, Phahonyothin Rd., Klong 1, Klong Luang, Pathumthani 12120

Email: supaporw@mtec.or.th Tel. +66 2564 6500 ext. 4042

Abstract This work introduces the use of AC-impedance spectroscopy (AC-IS) as a characterizing tool for cement-based applications. A thorough examination of the AC electrical behavior vs. microstructure relationships in cement-based composites has proven that AC-IS is sensitive to the presence of discontinuous conducting fibers in the composites. Below percolation, composite theory describes the relationship between the composite conductivity and filler volume fraction in terms of the intrinsic conductivity, which is a function of shape, orientation and electrical conductivity of filler. Such relationship can be related to the resistance values derived from AC-IS measurements to establish a foundation for AC-IS electrical characterization. Deviation from the predicted behavior is thus an indication of inhomogeneity and/or preferred orientation of the fibers. Some microstructural characterizations enabled by the AC-IS technique mentioned in the present work include after-processed fiber length (with known fiber volume fraction, and vice versa), fiber aggregation, and degree of fiber percolation. Good agreement with other characterization techniques confirms the validity of the AC-IS technique. Besides microstructural characterizations, AC-IS can potentially be used to characterize pozzolanic activity (i.e., the ability to provide SiO_2 to react with $Ca(OH)_2$ from cement hydration to form a strengthening product in cement-based materials) of rice husk ash (RHA), the most promising cement-replacement material for Thailand. Based on the rate of conductivity drop after the RHA is mixed with $Ca(OH)_2$ and water, pozzolanic activity characterization by AC-IS shows very good agreement with the amorphous SiO_2 content of the RHA studied.

1 Introduction

Cement is a technologically important ceramic material. Its exceptional compressive strength has manifested itself in the rise of countless skyscrapers worldwide. However, just as many other ceramics, the lack of toughness makes cement-based materials susceptible to catastrophic failures in the presence of flaws. The major goal in developing fiber-reinforced cement-based composites (FRCs) is thus to toughen the otherwise brittle cement matrix and/or to enhance its tensile and flexural strengths by fiber incorporation. The use of discontinuous or short fibers (e.g., short carbon or steel fibers) in reinforced cement-based composites has gain more attention compared to continuous fibers probably owing to its inexpensive fabrication, design flexibility, and relatively more isotropic properties.

Knowledge about the composite microstructure is crucial for the design of fabrication process and material selection in order to meet the desired properties. To that end, electrical characterization is advantageous compared to other microstructural characterization techniques (e.g., microscopy and image analysis) because it is non-destructive and not time-consuming. Furthermore, the method measures the response of the composite as a whole, not just a number of cross-sections assumed to statistically represent it, as in the microscopy and image analysis method.

AC-Impedance Spectroscopy (AC-IS) is a multi-frequency alternating current conductivity/impedance approach. It was originally developed to study electrochemical, interfacial, and corrosion reactions in systems with aqueous electrolytes [1], and later found applications in solid-state systems, such as solid electrolytes [2], electroceramics [3], and cement-based materials [4]. For electrochemical applications, the frequency range of interest over which most electrochemical/interfacial phenomena occur is typically in the sub-mHz to kH range. For solid-state applications, since it is more concerned with the bulk properties, the frequency range of interest is typically in the Hz to MHz range.

AC-IS applies a fixed low-amplitude AC excitation over a range of frequencies to the sample in order to measure its current response. The measured response, both magnitude and phase change, is typically represented in a Nyquist plot (negative imaginary impedance, -Im(Z), vs. real impedance, Re(Z)), with frequency increasing from right to left. The diameters of the semicircles or arcs in the plot represent the resistances of different components in the composite microstructure. Fig. 1. presents a typical Nyquist plot for ordinary Portland cement (OPC) with and without 0.06 vol% carbon fibers (18 mm initial length × 0.18 μm diameter) at 7 days of hydration. Darken points are frequency markers; the numbers labeled are log of frequency in Hz. The incomplete rightmost arc is associated with the electrochemical response of the external steel electrodes used to make measurements. This so-called electrode arc is present for both the plain OPC and the composite. To the left of the electrode arc are two bulk arcs for the composite, and a single bulk arc for the plain OPC. For each of the composite and the plain OPC, the low frequency cusp (intersection between the low-frequency bulk arc and the electrode arc) matches with their respective 4-point DC resistances, R_{DC}, shown as solid symbols on the Re(Z) axis. This suggests that the low-frequency resistance of the composite (~2500 Ω in Fig. 1.), whether measured by AC-IS or 4-point DC, remains unchanged from that of the unreinforced OPC, despite the addition of fibers.

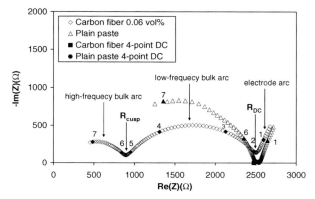

Fig. 1. Typical Nyquist plot for OPC with and without 0.06 vol% carbon fibers at 7 days of hydration. Also shown in the Re(Z) axis are the 4-pt DC resistances. (Reproduced from [5].)

Torrents et al. [6, 7] explained the origin of the dual bulk arcs in FRCs with short conducting fibers by the "frequency-switchable coating model". Given low fiber loadings (i.e., in the dilute regime), at DC and low AC frequencies, the conducting fibers behave as if they were insulating due to an electrochemical double-layer on the fiber interfaces that electrically isolates them from the matrix. Thus, at a small volume fraction, these insulating fibers will cause negligible change to the DC or low AC resistance of the composites [8]. At higher AC frequencies, displacement currents cause these double layers to short out, rendering the fibers conducting. This is reflected by the noticeable decrease in resistance of the composite at high frequencies, R_{cusp} (~900 Ω in Fig. 1.), when the conducting fibers serve as short-circuit paths throughout the composite. The model was later developed into a more quantitative "universal equivalent circuit model" to describe the impedance response of composites with conducting particulates or discontinuous fibers in the dilute regime [9].

The resistances derived from AC-IS can be used to characterize the composite microstructure based on the intrinsic conductivity approach. Assuming completely random distribution of fibers in the dilute regime, the conductivity of a composite, σ_c, containing fibers of conductivity, σ_f, and volume fraction, ϕ, in a matrix of conductivity, σ_m, is described by:

$$\frac{\sigma_c}{\sigma_m} = \frac{R_{DC}}{R_{cusp}} = 1 + [\sigma]_\Delta \phi + 0\phi^2; \quad \Delta \equiv \frac{\sigma_f}{\sigma_m} \quad (1)$$

The method treats each particle geometry as having an intrinsic conductivity, which is the first-order coefficient of ϕ in the above equation. Higher order terms are neglected assuming the particle loading remains in the dilute regime. For right-cylindrical fibers, the intrinsic conductivity, $[\sigma]_\Delta$, can be calculated from the fiber aspect ratio (AR or length/diameter) and Δ, which is the ratio of the fiber conductivity, σ_f, to that of the matrix. For highly conducting fibers (e.g., carbon or steel) where $\Delta \to \infty$, the intrinsic conductivity is expressed by the modified Fixman equation [8]:

$$[\sigma]_\infty = \frac{1}{3}\left(\frac{2AR^2}{3\ln(4AR)-7}+4\right) \tag{2}$$

However, for highly insulating fibers ($\Delta \to 0$) with AR > 10, the intrinsic conductivity, $[\sigma]_0$, is approximately -5/3, regardless of length. With such low volume fraction as 0.06%, this explains the negligible increase in R_{DC} when the carbon fibers behave as if insulating (i.e., at low frequencies) in Fig. 1. Conducting fibers with an aspect ratio of 100, for example, would result in $[\sigma]_\infty = 609$. Even at low fiber loadings, this still leads to a noticeable reduction in the high-frequency resistance (R_{cusp}) compared to R_{DC} of the unreinforced matrix, as evident in Fig. 1.

Equations (1) and (2) and their relations to the resistances values measured by AC-IS establish a foundation for electrical characterization by AC-IS. Any deviation from the predicted behavior is thus an indication of inhomogeneity or preferred orientation of the fibers. This work is, more or less, a review of the author's past and present involvements on the use of AC-IS as a characterizing tool for cement-based composites. The AC-IS characterizations covered in this work include determination of after-processed fiber length (with known fiber volume fraction, and vice versa), fiber aggregation, degree of fiber percolation, and pozzolanic activity evaluation. These topics will be presented separately in the following sections. The author was involved with the first three topics while pursuing a doctoral degree in Material Sciences and Engineering at Northwestern University, IL, USA. The last topic is the author's recent work involving cement-based materials while serving as a researcher at the National Metal and Materials Technology Center, Thailand.

2 Volume Fraction and Fiber Length Determination

For a known fiber volume fraction, AC-IS characterization can be employed to determine the fiber length, and vice versa. The former is especially useful for brittle fibers that are prone to breakage during processing such as carbon fibers. Using Eq. (1), the known fiber volume fraction, and the two resistance values measured by AC-IS, one can determine the fiber effective intrinsic conductivity. Eq. (2)

then allows a calculation of the aspect ratio and hence the after-processed fiber length, given that the fiber diameter remains unaltered after processing.

Hixson et al. [8] applied the method to extruded FRC specimen with 0.5 vol% carbon fiber. From the measured R_{DC}/R_{cusp} of 3.6 (after adjusting for fiber alignment in the extruded direction), and the fiber diameter of 10 μm, the after-processed fiber length was determined to be ~0.9 mm. This is in good agreement with the result from optical image analysis, after the cement matrix was dissolved by nitric acid and the collected fiber length distribution was measured. The median fiber length was 0.75 mm, with 70% being between 0.36 mm and 1.5 mm.

3 Fiber Aggregation

An AC-IS characterization method to study fiber dispersion in short FRCs has been established by Woo et al. [10]. Three dispersion issues, including fiber orientation, coarse-scale segregation, and local aggregation (clumping) were investigated. Fiber aggregation, in particular, was evaluated in terms of a "dispersion factor." It compares the electrical response of composite with clumping (Eq. (3)) to the predicted behavior (Eq. (1)), which assumes randomly distributed, randomly oriented (i.e., well-dispersed) fibers.

$$\frac{\sigma_c}{\sigma_m} = 1 + [\sigma]_\infty \phi' + \Sigma[\sigma]_{\infty,j}\phi_j \tag{3}$$

ϕ' is the volume fraction of randomly dispersed fibers of intrinsic conductivity, $[\sigma]_\infty$, and ϕ_j is the volume fraction of the j-th fiber clump of intrinsic conductivity, $[\sigma]_{\infty,j}$. The overall volume fraction of fibers, ϕ, can be expressed as $\phi = \phi' + \Sigma\phi_j$. Dividing Eq. (3) by Eq. (1) yield the dispersion factor (DF), given by:

$$DF = \left[\frac{\left(\dfrac{\sigma_c}{\sigma_m}\right)_{measured} - 1}{\left(\dfrac{\sigma_c}{\sigma_m}\right)_{predicted} - 1}\right] = \frac{\phi'}{\phi} + \frac{\Sigma[\sigma]_{\infty,j}\phi_j}{[\sigma]_\infty\phi} \tag{4}$$

The first term of the rightmost expression represents the fraction of well-dispersed fibers. The second term represents the contribution to the composite conductivity from fiber clumps. Although small, the second term is non-zero, depending on the exact nature of fiber aggregation. Therefore, the DF serves only as the upper limit for the fraction of well-dispersed fibers. However, for a composite with perfectly dispersed fibers, the second term must go to zero and $\phi' = \phi$, yielding DF = 1. Therefore, a DF close to one indicates good dispersion or negligible fiber aggregation, whereas a DF less than one indicates a tendency for fiber clumping. A smaller DF thus signifies poorer fiber dispersion. Experimentally, the DF is calculated from the measured resistances, the intrinsic conductivity, $[\sigma]_\infty$, and the overall fiber volume fraction, ϕ, as followed:

$$DF = \frac{\left(\dfrac{R_{DC}}{R_{cusp}}\right)_{ave} - 1}{[\sigma]_\infty \phi} \qquad (5)$$

The term $(R_{DC}/R_{cusp})_{ave}$ is the average of (R_{DC}/R_{cusp}) measured from three orthogonal directions to eliminate the effect of preferred fiber alignment.

The method was applied to four FRC systems with different admixtures and different fiber surface treatments (in order to obtain different degrees of fiber aggregation). The compositions, mixing procedures, and electroding of the composites were mentioned in details elsewhere [11]. The four composites contained similar loadings (~0.4 vol%) of carbon fibers, whose initial dimensions were 5 mm in length and 15 μm in diameter. However, since processing of the brittle carbon fibers can lead to a distribution of fiber aspect ratio in the actual composites, an estimation of the intrinsic conductivity, $[\sigma]_\infty$, in Eq. (5) for each system was obtained from the actual after-processed fiber length distribution and used to determine the DF. Three cubic replicates were measured for each of the four composite systems. For each replicate, AC-IS measurements were made along each of the three orthogonal directions of the cube.

Figure 2 shows the resulting DFs from the AC-IS measurements for the four composites. Three DFs, one for each of the three replicates (open symbols), along with an average DF (solid symbols), are shown for each type of composites. The error bars were calculated based on the 5% experimental uncertainty in the interelectrode spacing for the AC-IS measurements and the variation in the DFs among the three replicates. A set of representative cross-sections, depicting different degrees of fiber dispersion for each system, is shown in Fig. 3 for comparison.

The A replicate appears to be relatively homogeneous, while the B replicate shows only a few fiber-rich regions. These observations are consistent with the high DFs derived from AC-IS for the two systems in Fig. 2. Extensive networks of voids surrounded by fiber-rich regions are evident in the C replicate. Very large fiber pockets are observed, along with a few fiber-rich areas, in the D replicate. These findings, again, are in good agreement with the low DFs for both systems in Fig. 2. It should be noted, however, that the different degrees of fiber dispersion between system A and B, and between system C and D, are not sufficient for AC-IS to detect the difference in the bulk electrical conductivity, resulting in similar DFs. Nevertheless, the agreement between the AC-IS-derived dispersion factors and the optical micrographs has demonstrated that AC-IS, combined with an intrinsic conductivity approach and a knowledge of the actual fiber length distribution, can be used to characterize fiber aggregation in short conductive FRCs.

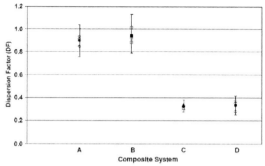

Fig. 2. Dispersion factors for the four composite systems.

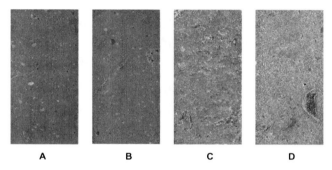

Fig. 3. Representative optical micrographs showing different degrees of fiber dispersion in the four composite systems.

4 Fiber Percolation

To investigate the potential use of AC-IS to characterize fiber percolation, plain cement paste with and without 0.75 vol% or 1 vol% multi-walled carbon nanotubes (MWCNTs, initially 10-20 nm × 1-2 μm length) were fabricated. Details on the mixing procedures and electroding were given by Wansom et al. [5]. At 1 day of hydration, AC-IS measurements were made on three replicates for each composition to test for reproducibility. Four-point DC measurements were also performed to extract the DC resistances of the composites.

The AC-IS responses of the plain cement paste and the two composites are shown in Fig. 4. A direct mapping of features between MWCNT-FRC spectra and microfiber FRC spectra (Fig. 1.) can be realized, except for the depressed low-frequency bulk arc, in contrast to a clear arc-shape in Fig. 1. The intersection of the low-frequency bulk arc and the electrode arc of the composite still coincides with its DC resistance, as is the case for microfiber FRC. However, a marked decrease in the DC resistance of the MWCNT-FRC, R_{DC}(FRC), vs. that of the unreinforced matrix, R_{DC}(matrix), is a major difference from the behavior of microfiber FRC, where the composite DC resistance remains essentially unchanged compared to that of the matrix.

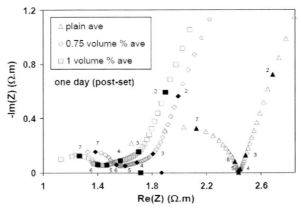

Fig. 4. Nyquist plots for plain cement paste and the two composites (with 0.75 vol% and 1 vol% MWCNTs), averaged from three replicates for each composition. Note the corresponding 4-point DC resistances on the Re(Z) axis.

The shift in DC resistance of the MWCNT-FRC from that of the matrix is a result of percolation of the carbon nanotubes. At DC or low AC frequencies, current now travels through the composite via both the matrix and the percolation paths, resulting in a DC resistance lower than that of the matrix alone. Based on the parallel fiber percolation and matrix paths, the measured R_{DC}(FRC) and R_{DC}(matrix), and the resistance of an individual nano-strand, one can estimate the number of such percolating nano-strand spanning the electrodes as being 850 and 1270 strands for the 0.75 vol% and 1 vol% MWCNT-FRCs, respectively. These estimations are consistent with a more extensive percolation network expected in specimen with higher loading of MWCNTs.

5 Pozzolanic Activity of Rice Husk ash

Pozzolanic activity refers to the material's ability to provide reactive SiO_2 and/or Al_2O_3 to react with $Ca(OH)_2$ from cement hydration to form a strengthening product in cement-based materials. To investigate the potential of AC-IS to characterize pozzolanic activity of rice husk ash (RHA), various RHAs from different incinerating conditions were mixed with $Ca(OH)_2$ and water, following the mix proportions suggested by McCarter and Tran [12]. The changes in conductivity of the different mixtures were monitored by continuous AC-IS measurements from 15 min to ~52 h after the addition of water. The method relies on the consumption of Ca^{2+} and OH^- by the pozzolanic reaction, which leads to the decrease in conductivity of the mixture as the reaction proceeds over time. Hence, a faster conductivity drop would indicate higher pozzolanic activity. The conductivity values at later times were normalized by the earliest (t ~ 5 min) conductivity to account for the conductivity contributions from other soluble salts and unburnt carbon in the RHAs. The normalized conductivity change is shown in Fig. 5.

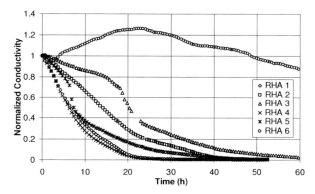

Fig. 5. Normalized conductivity vs. time for the different RHA + Ca(OH)$_2$ pastes.

Table 1. Agreement between the amorphous SiO$_2$ content and the rate of normalized conductivity change during the first 24 h of the reaction

Sample	Amorphous SiO$_2$ (wt%)	$d\sigma/dt_{24h}$ (h^{-1})
RHA 1	83.81	-4.02×10^{-2}
RHA 2	82.27	-3.62×10^{-2}
RHA 3	62.38	-2.36×10^{-2}
RHA 4	90.94	-4.03×10^{-2}
RHA 5	85.38	-3.94×10^{-2}
RHA 6	38.77	1.51×10^{-2}

The rate of the normalized conductivity change during the first 24 h of the reaction, $d\sigma/dt_{24h}$, was found to be in very good agreement with the amorphous (or reactive) SiO$_2$ content of the different RHAs, as shown in Table 1 [13]. The correlation coefficient between the two variables for the set of RHAs investigated is as high as -0.97. This confirms the sensitivity of the AC-IS characterization to the amorphous SiO$_2$ content, which is one of the most important factors known to influence the pozzolanic activity of cement-replacement materials.

6 Conclusions

AC-impedance spectroscopy (AC-IS), coupled with an intrinsic conductivity approach, show potential to be used as a reliable characterizing tool for various microstructural aspects of short fiber-reinforced cement-based composites. Some success in using the technique to characterize the after-process fiber length, fiber aggregation, fiber percolation, and pozzolanic activity of rice husk ash were reviewed in details. Good agreement with other characterization techniques confirms the validity of the AC-IS technique.

Acknowledgments

The author would like to thank Professor Thomas O. Mason and Dr. Leta Y. Woo for their contributions to the majority of this work. Siam Research and Innovation Co., Ltd. is acknowledged for the financial support and for providing samples analyzed in the last part of the work.

7 References

[1] Macdonald JR (1987) Impedance spectroscopy: Emphasizing solid materials and systems. John Wiley and Sons, Inc., New York
[2] Bauerle JE (1969) Study of solid electrolyte polarization by a complex admittance method. J Phys Chem Solids 30:2657–2670
[3] Irvine JTS, Sinclair DC, West AR (1990) Electroceramics: Characterization by impedance spectroscopy. Adv Mater 2(3):132–138
[4] Christensen BJ, Coverdale RT, Olson RA, Ford SJ, Garboczi EJ, Jennings HM, Mason TO (1994) Impedance spectroscopy of hydrating cement-based materials: measurement, interpretation, and application. J Am Ceram Soc 77(11):2789–2804
[5] Wansom S, Kidner NJ, Woo LY, Mason TO (2006) AC-impedance response of multi-walled carbon nanotube/cement composites. Cem Concr Comp 28(6):509–519
[6] Torrents JM, Mason TO, Peled A, Shah SP, Garboczi EJ (2001) Analysis of the impedance spectra of short conductive fiber-reinforced composites. J Mater Sci 36(16):4003–12
[7] Torrents JM, Mason TO, Garboczi EJ (2000) Impedance spectra of fiber reinforced cement-based composites: a modeling approach. Cem Concr Res 30(4):585–92
[8] Hixson AD, Woo LY, Campo MA, Mason TO, Garboczi EJ (2001) Intrinsic conductivity of short conductive fibers in composites. J Electroceram 7(3):189–195
[9] Woo LY, Wansom S, Hixson AD Campo MA, Mason TO (2003) A universal equivalent circuit model for the impedance response of composites. J Mater Sci 38(10):2265–2270
[10] Woo LY, Wansom S, Ozyurt N, Mu B, Shah SP, and Mason TO (2005) Characterizing fiber dispersion in cement composites using ac-impedance spectroscopy. Cem Concr Comp 27(6):627–636
[11] Wansom S (2006) Electrical impedance characterization of ceramic-matrix composites with various inclusions. Ph. D. Dissertation. Northwestern University, Evanston, IL, USA
[12] McCarter WJ, Tran D (1996) Monitoring pozzolanic activity by direct activation with calcium hydroxide. Const Build Mater 10:179-184
[13] Wansom S, Janjaturaphan S, Sinthupinyo S (2009) Pozzolanic activity of rice husk ash: comparison of various electrical methods. J Met Mater and Miner 19(2):1-7

On the use of indentation technique as an effective method for characterising starch-based food gels

C. Gamonpilas[1*], M.N. Charalambides[2] and J.G. Williams[2]

[1]National Metal and Materials Technology Center, National Science and Technology Development Agency, Pathumthani 12120, Thailand

[2]Department of Mechanical Engineering, Imperial College London, London SW7 2AZ, UK

*Corresponding e-mail: chaiwutg@mtec.or.th; Tel. 02-5646500 ext. 4447; Fax: 02-5646446

Abstract The use of indentation technique to characterise the mechanical properties of food materials was assessed in this work. Two different types of starch gels were used as testing materials. Indentation tests were performed on these materials which revealed a rate independent load-deflection response. An inverse analysis based on the Marquardt-Levenberg optimisation algorithm and finite element analysis was implemented in order to derive the stress-strain behaviour from the experimentally obtained indentation data. The inverse predictions for the stress-strain curves were in good agreement with the direct measurements from uniaxial compression and shear tests up to high strain values. The feasibility and effectiveness of the method was demonstrated for both stiff and compliant gels with initial moduli ranging from a very small 60 Pa to 55 kPa. Thus the indentation characterisation method was proven as a powerful, fast and efficient way of evaluating and/or monitoring the behaviour of food materials such as starch gels.

1 Introduction

Large deformation properties of food materials are very important for handling, processing as well as in masticating. Obtaining intrinsic mechanical properties of such materials is a challenging task due to their soft nature and significant variation between different batches [1-2]. A range of tests are available for characterising the mechanical behaviour of soft materials such as gels or foods. These tests can be categorised into those with a uniform strain distribution such as the uniaxial compression and tension tests, and those with non-uniform strain deformation such as the indentation test. The former tests are well-established for conventional structural materials because the measured loads and displacements can be converted easily into the stress–strain response using simple expressions. However, such tests can be difficult to apply to soft foods due to the low stiffness and constraints in size and shape. On the other hand, the indentation test offers many advantages because it is quick and simple and does not require samples with specific shape requirements. Furthermore, the indentation response is not greatly affected by friction between sample and loading platens, as in the case of uniaxial compression [3]. The test can also be performed on a large scale using high throughput screening techniques where large numbers of different product formulations/recipes can be automatically tested using high speed robotic indenters on small samples moving on conveyor belts. However the mechanical behaviour of food is usually very non-linear, hence there are no theories that can be applied to relate the indentation load–displacement response to the stress–strain properties [4, 5]. As a consequence, there is a need for converting indentation load–depth data for non linear materials into the fundamental material properties using an inverse parameter identification technique so that sensory assessment of foods can be improved.

Inverse analysis techniques have been extensively use in various engineering applications including signal processing, inertial navigation, radar tracking, manufacturing and many other aspects [6]. However, the use of such techniques to determine the material properties from indentation is comparatively new with little work published regarding its application in conjunction with the indentation test. Until recently, the inverse approach and instrumented indentation technique have been used to successfully evaluate the material parameters of functionally graded materials [7], elastoplastic [8] and viscoelastic materials [3]. Thus, the aim of this work is to use the inverse analysis with experimental data obtained from indentation tests to derive the stress-strain characteristics of food materials. Starch gels made with varying concentrations were used as testing materials. The results obtained from the inverse calculations were validated by independent compression and shear experiments.

2 Materials and preparation

Two different types of starch gel samples prepared using modified sago and modified maize starch powders were used in this work. These gels were made as described in [9] and are arbitrarily called 2-34 and 2-12, respectively. Various powder concentrations were studied e.g. 5, 10, 15 and 20% w/w (ratio of weight of powder to weight of water) for the 2-34 and 10, 15 and 20% w/w for the 2-12. It is worth noting that 2-34 (5%) and 2-12 (10%) gels are non-self supporting i.e. very soft gel.

3 Experiments

All experiments were conducted at 21°C and 50% relative humidity. Indentation tests were performed on an Instron 5543 testing machine using a spherical indenter of diameter 5.90 mm at the three loading speeds of 0.5, 5 and 50 mm/min. In order to validate the proposed inverse analysis methodology (see later section), uniaxial compression tests were also performed on all self-supporting gels by moving the Instron crosshead at constant strain rates of 0.1, 1 and 10 /min. Silicon oil was used throughout the tests in order to minimise friction between the compression platens and the sample and to eliminate sample height effects [1]. Furthermore, storage (G') and loss (G'') moduli as well as the shear stress-strain behaviour of non-self supporting gels were measured using a parallel plate geometry with a diameter of 40 mm and a gap of 1 mm (AR2000ex, TA instruments). These gels were too weak to be tested in compression and this is the reason why shear experiments were performed instead. A frequency sweep (0.1-63 rad/s) test within the linear viscoelastic range, i.e. with 2% strain amplitude, was conducted. In addition, a constant shear rate test was also performed at a rate of 0.05/s until rupture of the gel was obtained.

4 Hyperelastic Constitutive Model

It will be shown later that the stress-strain behaviour of starch gels can be described by the hyperelastic constitutive model based on Ogden solution. Under a uniaxial deformation state, the true stress, σ, in the Ogden solution is defined as

$$\sigma = \frac{2\mu}{\alpha}\left(\lambda^{\alpha} - \lambda^{-\frac{1}{2}\alpha}\right) \tag{1}$$

where λ is the stretch ratio and $\lambda = \exp(\varepsilon)$ where ε is the true strain, μ is the initial shear modulus ($E=3\mu$) and α is the Ogden constant [10]. In addition, the shear stress, τ, under a simple shear condition is related to λ as shown below

$$\tau = \frac{2\mu}{\alpha}\left(\frac{\lambda^{\alpha+1}}{\lambda^{2}+1} - \frac{\lambda^{-\alpha+1}}{\lambda^{2}+1}\right) \tag{2}$$

where λ is related to the shear strain, γ, through $\lambda = \frac{\gamma}{2} + \frac{\sqrt{\gamma^{2}+4}}{2}$ [11].

5 Indentation Forward Predictions

The objective of the forward prediction was to predict the indentation force-displacement response using the constitutive properties as measured from the compression tests. This step was necessarily performed prior to implementing the inverse analysis in order to demonstrate that the indentation response of these materials can be accurately predicted if their stress-strain properties are known. Since there is no analytical solution for the indentation of non linear materials, the predictions were obtained from finite element (FE) simulations of the indentation tests using the commercial finite element code ABAQUS [10].

The problem was modelled as a cylinder being indented by a rigid spherical indenter due to a geometric and loading symmetry about the axis of the indenter. A very fine mesh was used in the vicinity of contact region due to the large deformation of the region. A gradually coarser mesh was applied away from the contact region in order to reduce the total number of degrees of freedom and computational time. A mesh sensitivity study was performed and the final mesh used is shown in Fig. 1. The contact between the indenter and the specimen was assumed to be frictionless. During the analysis, the rigid indenter was loaded by means of a downward displacement along the y axis. The loading curve was obtained directly from the normal reaction force on the rigid indenter as a function of the vertical displacement of the indenter.

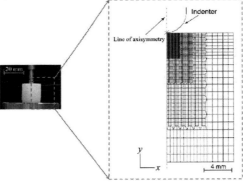

FIG. 1. Illustration of finite element analysis of an indentation problem

6 Indentation Inverse Predictions

The proposed inverse analysis utilises the Levenberg-Marquardt (LM) method which is described extensively in the work of Schnur and Zabaras [12]. Its use within the context of this work is illustrated by the flow chart shown in Fig. 2. Essentially, it processes the experimental data and attempts to obtain the best estimates for unknown state variables based on least-squares theory.

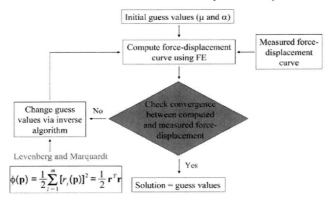

FIG. 2. Illustration of an inverse analysis procedure

The theory of the LM method is based on minimising an error function, Φ, with respect to the parameter, $\mathbf{p}$, as

$$\Phi(\mathbf{p}) = \frac{1}{2}\sum_{i=1}^{m}[r_i(\mathbf{p})]^2 = \frac{1}{2}\mathbf{r}^T\mathbf{r} \tag{3}$$

where $\mathbf{p}$ is a vector which contains the unknown parameters and m is the number of measurements. The vector $\mathbf{r}$ is defined as

$$\mathbf{r} = \mathbf{p}_i - \hat{\mathbf{p}} \tag{4}$$

where $\mathbf{p}_i$ and $\hat{\mathbf{p}}$ are the numerically generated and measured indentation load, respectively, at the specified indenter displacement values.

From Eq. (1), two parameters (μ and α) are required to describe the non-linear elastic behaviour. Starting with initial guess values of μ and α, FE analysis of the indentation problem was performed to obtain the indentation loading response, $\mathbf{p}_i$. This resulting curve is subsequently compared with the measured one, $\hat{\mathbf{p}}$, through Eq. (3). The problem is iteratively solved through the inverse procedure until the solution converges. The computations were conducted using ABAQUS [10].

7 Results and discussion

7.1 Uniaxial compression response

Fig. 3 shows the typical compression behaviour of 2-34 (10%) and 2-12 (20%) gels at three constant strain rates of 0.1, 1 and 10/min. Independent loading-unloading tests on these gels also revealed reversible response although results are not shown here. Therefore, the gels can be said to deform in a non-linear elastic manner and are rate independent. Such findings are in agreeable with the previous work on starch gels [13, 14]. The compression response of both 2-34 and 2-12 starch gels at various concentrations were calibrated with the Ogden model given by Eq. (1), using a Solver function in Excel. Fig. 4 compares the predictions from the model with the experimental data. The results show excellent agreement between the model and the experimental data, confirming the suitability of the model used. Values of the material constants, μ and α, for each material are tabulated in Table 1.

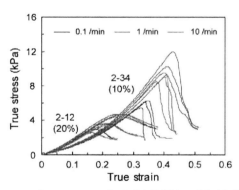

FIG. 3. Compression response for 2-34 (10%) and 2-12 (20%) gels.

TABLE 1. Material parameters of 2-34 and 2-12 gels at various concentrations.

Ogden	2-34 (10%)	2-34 (15%)	2-34 (20%)	2-12(15%)	2-12(20%)
µ (kPa)	3.3	22.9	56.5	2.1	5.0
α	-4.9	-4.1	-3.7	-4.9	-2.9

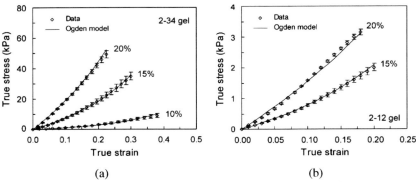

FIG. 4. Compression response of (a) 2-34 and (b) 2-12 gels at various concentrations (%w/w) fitted with the Ogden model.

7.2 Indentation response

A typical indentation load-depth response of 2-34 (10%) and 2-12 (20%) gels is shown in Fig. 5. In agreement with their compression behaviour, the indentation response of these gels is found to be speed independent.

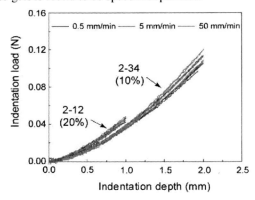

FIG. 5. Indentation response for 2-34 (10%) and 2-12 (20%) gels.

7.3 Indentation forward predictions

Finite element predictions of the indentation response are compared to experimental data in Figs. 6(a) and 6(b) for 2-34 and 2-12 gels, respectively. The agreement is excellent which justifies the effectiveness of the numerical models and the assumption of the hyperelastic model based on Ogden to characterise these starch gels. These accurate forward predictions suggest that an inverse analysis using indentation data might lead to the stress-strain properties of these starch gels.

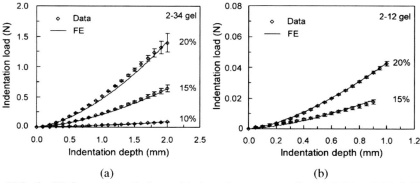

FIG. 6. FE forward predictions of indentation response for (a) 2-34 and (b) 2-12 gels at various concentrations.

7.4 Indentation inverse predictions

Using the average indentation force-displacement, the constitutive behaviour of both starch gels was predicted using the inverse methodology outlined in Section 6. The predicted material parameters for 2-34 and 2-12 gels at various concentrations are summarised in Table 2. The percentage difference between the parameter values in Tables 1 and 2 are given in Table 3. For all gels studied, the inverse analysis predicted the values of μ and α within maximum differences of 8.7% and 26.1%, respectively, as compared to the compression values. The reason for the larger difference seen in the values of α is due to the low sensitivity of the stress-strain curve on this parameter when the latter is in the range of -3 to -5. Furthermore, it is clearly demonstrated in Fig. 7 that the inverse predictions agree very well with the compression data for all materials studied, suggesting that the inverse methodology is a powerful tool for obtaining the constitutive behaviour of gels from its available indentation data.

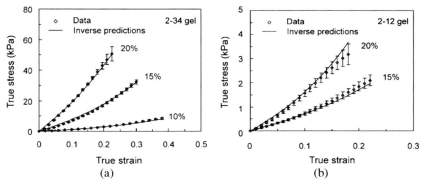

FIG. 7. Inverse analysis predictions of stress-strain response from indentation data for (a) 2-34 and (b) 2-12 at various concentrations.

TABLE 2. Inverse predictions of material parameters of 2-34 and 2-12 gels at various powder concentrations.

Ogden	2-34 (10%)	2-34 (15%)	2-34 (20%)	2-12(15%)	2-12(20%)
μ (kPa)	3.1	21.9	55.3	2.2	4.7
α	-6.2	-5.1	-4.4	-6.6	-3.7

TABLE 3. Percentage difference between material parameters predicted from compression tests (Table 1) and inverse analysis of indentation test data (Table 2).

Ogden	2-34 (10%)	2-34 (15%)	2-34 (20%)	2-12(15%)	2-12(20%)
μ (kPa)	8.7%	4.4%	2.2%	-4.2%	7.5%
α	-20.5%	-19.7%	-15.5%	-26.1%	-21.5%

7.5 Non-self supporting gels

A challenging task in this work was how to characterise the two non-self supporting gels using indentation data. In order to solve this problem, indentation tests were performed within the mould containing the gel, as shown in Fig. 8(a), since samples were not able to support themselves when left standing alone. A typical indentation response is shown in Fig. 8(b) for the 2-34 (5%) gel. It is apparent that the behaviour is still rate-independent and similar to that of the self-supporting gels. Some fluctuations in the measurements were obtained due to the dynamic effect from the single screw mechanism of the testing machine.

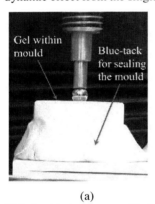
(a)

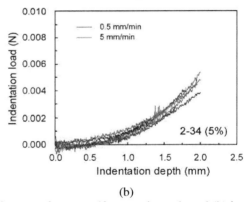
(b)

FIG. 8. (a) Apparatus of indentation tests for non-self supporting gels and (b) indentation response of 2-34 (5%) performed within the mould.

In order to perform the inverse analysis on the indentation data of non-self supporting gels, the FE model described in Section 5 needs to be modified by prescribing an additional constraint along the boundary between the sample and the mould. The indentation data of non-self supporting gels were also 'smoothed' by fitting with a simple quadratic equation in order to improve the predictive accuracy from the inverse analysis. Using the smoothed response, the inverse predictions were subsequently performed on both gels to obtain their μ and α parameters. The results for non-supporting gels are summarised in Table 4.

TABLE 4. Inverse predictions of material parameters of non-self supporting gels.

Ogden	2-34 (5%)	2-12 (10%)
μ (Pa)	62.70	223.81
α	-9.86	-9.10

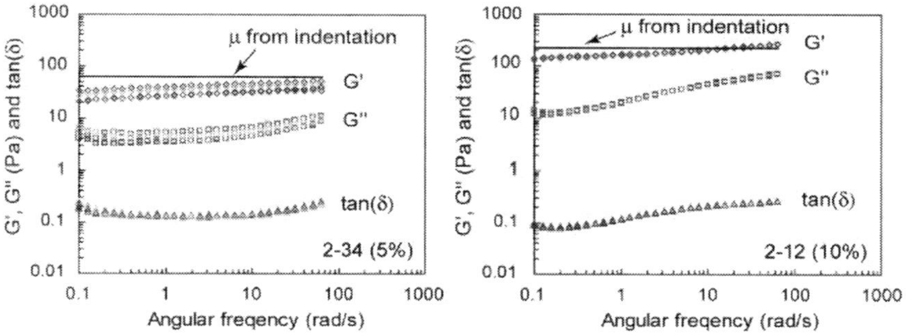

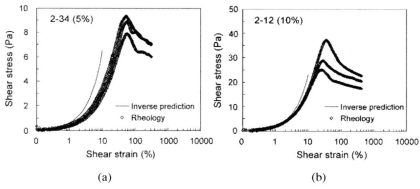

FIG 9. G', G'' and $\tan\delta$ spectra as a function of frequency for (a) 2-34 (5%) and (b) 2-12 (10%).

FIG. 10. Rheological behaviour of 2-34 (5%) and 2-12 (10%) in comparison with those predicted from inverse predictions of their indentation data.

As uniaxial compression tests were not possible for the non-self supporting gels, rheometric shear tests were performed instead, in an effort to validate the inverse predictions. The curves of G', G'' and $\tan\delta$ as a function of frequency are shown in Figs. 9(a) and 9(b) for 2-34 (5%) and 2-12 (10%), respectively. It can be seen that G' is much larger than G'' in the entire frequency domain and $\tan\delta$ is small and almost constant. These spectra are typical of those usually observed for gels. When increasing the shear strain by applying a very low steady shear rate to the gel, the shear stress started to increase proportionally up to the shear strain of about 1% for both gels. Beyond this point, the shear stress increased more rapidly with the shear strain and dropped sharply at shear strain of approximately 0.4 for the 2-34 (5%) and 0.2 for the 2-12 (10%) gels (see Fig. 10). These results are similar to the work on biopolymer weak gels by Michon and co-workers [15].

In addition, the μ values shown in Table 4 are plotted together with the storage moduli in Fig. 9. It is evident that the predicted μ values for both gels are in good agreement with the storage moduli obtained from the rheological. The predictions of the shear behaviour of both gels calculated from Eq. (2) are also shown in comparison with the rheological measurements in Fig. 10. Agreement between the two tests is reasonable for both materials. The discrepancies in the predictions may be due to experimental difficulties in performing both tests. On one hand, it could be the result of large experimental scatter in indentation data as the level of load measured was comparable with the dynamic effect of the machine. On the other hand, there could be some slippage occurring during the rheological measurements, hence a weaker response.

8 Conclusions

Mechanical/rheological behaviour of two starch gels with various starch/water weight ratios, forming either self supporting or non-self supporting gels have been obtained using indentation tests. A method based on finite element and inverse analysis techniques was developed to inversely predict the stress-strain behaviour of these gels from their indentation data. The inverse predictions were validated with the independent compression and shear data for self-supporting and non-self supporting gels, respectively. Excellent agreement was obtained for all of the gels studied.

9 Acknowledgements

Support for this work has been provided by ICI plc under the SRF scheme. ABAQUS was provided under academic license by HKS Inc., Providence, Rhode Island, USA.

References

[1] Charalambides MN, Goh SM, Wanigasooriya L et al (2005) Effect of friction on uniaxial compression of bread dough. J Mater Sci 40:3375-3381.
[2] Goh SM, Charalambides M, Williams JG (2005) On the mechanics of wire cutting of cheese. Eng Fract Mech 72:931-946.
[3] Goh SM, Charalambides MN, Williams JG (2004) Characterisation of non-linear viscoelastic materials via the indentation techniques. Rheolo Acta 44:47-54.
[4] Sneddon IN: The relation between load and penetration in the axisymmetric Boussinesq problem for a punch of arbitrary profile, Int J Eng Sci 3 (1965) 47–57.
[5] Sakai M, Time-dependent viscoelastic relation between load and penetration for an axisymmetric indenter, Phil Mag A 82 (2002) 1841-1849.
[6] Grewal MS, Andrews AP (1993) Kalman Filtering: Theory and practice, Prentice-Hall Inc., New Jersey.

[7]   Nakamura T, Wang T, Sampath S (2002) Determination of properties of graded materials by inverse analysis and instrumented indentation. Acta Mater 48:4293-4309.

[8]   Gamonpilas C, Busso EP (2007) Characterization of elastoplastic properties based on inverse analysis and finite element modelling of two separate indenters, J Eng Mater Tech 129:603-608.

[9]   Gamonpilas C, Charalambides MN, Williams JG (2010) Predicting the mechanical behaviour of starch gels through inverse analysis of indentation data, Appl Rheol 20:33283.

[10]   Abaqus version 6.4 (2004), Hibbitt Karlsson and Sorensen Inc, Providence, RI.

[11]   Treloar LRG (1975) The physics of rubber elasticity, Oxford University Press, UK.

[12]   Schnur DS, Zabaras N (1992) An inverse analysis for determining elastic material properties and a material interface. Int J Numer Methods in Eng 33:2039-2057.

[13]   Luyten H, Van Vliet T (1995) Fracture properties of starch gels and their rate dependency. J Texture Stud 26:281-298.

[14]   Ikeda S, Yabuzoe T, Takaya et al (2001) Effects of sugars on gelatinisation and retrogradation of corn starch. In: Barsby TL, Donald AM and Frazier PJ (ed) Starch: Advances in structure and function, RSC, Cambridge, UK 67-77.

[15]   Michon C, Chapuis C, Langendorff V et al (2004) Strain-hardening properties of physical weak gels of biopolymers. Food Hydrocolloid. 18:999-1005.

Photothermal Radiometry applied in nanoliter melted tellurium alloys

A. Cappella [1,2,*], **J.-L. Battaglia** [2], **V. Schick** [2], **A. Kusiak** [2], **C. Wiemer** [3], **M. Longo** [3], **B. Hay** [1]

[1] Scientific and Industrial Metrology Direction, Laboratoire National Optical Division, Bureau Nationale de Metrologie – LNE, Trappes, France

[2] Laboratoire TREFLE, UMR 8508, University of Bordeaux, 33405 Talence, France

[3] Laboratorio MDM, IMM-CNR, via C. Olivetti 2, 20041 Agrate Brianza, (Mi), Italy

Abstract

We report on thermal measurements of molten materials at the nanoliter scale. An experimental setup of Photothermal Radiometry (PTR), formerly developed for solid state measurements, has been adapted for this purpose. The material is a chalcogenide glass-type tellurium alloy, $Ge_2Sb_2Te_5$ (GST), amorphous at room temperature, and that becomes crystalline at 130°C. The same material, brought to its melting temperature T_m, about 600°C, becomes amorphous after rapid cooling. Since the liquid is the precursor phase of the amorphous state, its characterization is of paramount importance. Thin film PTR characterization was first performed in solid state by measuring the GST thermal conductivity evolution during the structural phase changing, from the amorphous phase to its crystalline phase. In order to characterize the melt at high temperature, a lightly Ge-doped Te alloy sample was secondly fabricated. This latter tellurium alloy melts at a lower temperature, (T_m~450°C, as for pure tellurium) than GST. A random lattice of hemispherical tellurium structures, 500 nm in radius, was grown by MOCVD technique on a thermally oxidized silicon substrate. The hemispheres were then embedded in a 500 nm SiO_2 protecting layer in order to prevent evaporation during the melting. A 30 nm cap layer of Pt was then evaporated on the SiO_2 as thermal transducer for the laser beam. Measurements have been performed from room temperature up to 650°C. SEM and XRD measurements performed after annealing, have shown that these samples withstood the thermal stress up to 300°C. At temperatures above 380°C some Te is still present in the hemispherical structures, but a part of it has reacted with Pt to form PtTe by migration through the SiO_2 matrix. Experiments carried out at temperatures below 300°C have shown an anomalous behaviour of the thermal contact resistance between the tellurium alloy and the oxide interface.

---

* To whom correspondences should be addressed: *andrea.cappella@lne.fr*

1 INTRODUCTION

Tellurium is a component element among many of the important II-VI pseudo-binary semiconducting alloys, as HgCdTe, HgZnTe, CdZnTe, or of the IV-VI chalcogenide alloys, similar to the ones that lie along the GeTe-Sb_2Te_3 pseudo-binary line. Applications for Te compounds range from infrared optics to electronics [1, 2]. Thin films of these materials show interesting properties, the most remarkable being the ability of rapid and reversible transitions between different structural phases. Consequently, these materials show important changes in its electrical resistance and its optical reflectivity, depending on the crystal phase. Heat is the driving process of this reversible switching. The crystallization is induced by heating the alloy at temperatures between the glass transition (T_g) and melting (T_m) temperatures. Fast-quenching from the melt switch-back the material in its amorphous phase. Among all the tellurium alloys, the $Ge_2Sb_2Te_5$ (GST) chalcogenide ternary compound is the most popular one, since it is characterized by good switching-speed and long-term stability [3]. This alloy is employed as memory element inside the well-known Rewritable DVDs and inside the next generation of non-volatile solid-state memories, the Phase Change Memories (PCMs) [4]. The GST alloy is stable at room temperature in the amorphous and hexagonal crystalline *hcp*-phase, and metastable in the face centered cubic *fcc*-phase [5]. The switching temperature, that is the glass-transition temperature, is around 130°C for the *fcc*-crystalline phase and around 350°C for *hcp*-crystalline phase, whereas the melting temperature is approximately 600°C. The electrical resistivity varies over several orders of magnitude, according to the phase [6]: in the amorphous state GST is an insulator (up to 1 Ω m), whereas in the crystalline state, it becomes conductive (10^{-4}–10^{-3} Ω m for the fcc crystalline and about 10^{-5} Ω m for the hcp crystalline). Astonishingly the GST thermal properties do not exhibit the same dramatic change during the switching [7, 8]. Values for the thermal conductivity are reported in literature ranging from 1.7 to 2 W/m/K for the *hcp* crystalline GST, and are around 0.2 W/m/K for the amorphous GST. Nevertheless, even if the variation is small, the thermal conductivity remains a sensitive parameter when simulating the rapid heat transfer inside PCM memories [9, 10]. Moreover, as the minimum size of the features of electronic devices continues to decrease, the small interfacial thermal resistance between layers, generally around 10^{-9}-10^{-8} K m^2 W^{-1}, becomes comparable with the thermal resistance of the film itself, and it is increasingly important in the heat transfer process [11]. Previous works have shown the *in situ* evolution of the GST thermal conductivity and GST thermal boundary resistance during the phase changing [12, 13], while, to our knowledge, no thermal data about the GST melting phase have been published. Nevertheless, the characterization of the melted GST is a really interesting matter since the melt is a key step in the amorphization process.

This paper describes the results of recent investigations particularly directed to extend the characterization of thermal properties of a chalcogenide materials to its melt phase by PhotoThermal Radiometry (PTR). Complementary information is provided by Scanning Electron Microscopy (SEM) and gracing incident X-Ray Diffraction (XRD). The PTR technique has been successfully implemented to characterize, *in situ*, the thermal properties of thin GST films below the melting temperature, from the amorphous phase to the crystalline one. Conversely special arrangements were needed to extend the measurement above the melting temperature of a nano-scaled material, because instantaneous evaporation at high temperatures. Actually, heating the samples up to 600°C and preserving their integrity during the long-lasting PTR measurement, are a really hard task. The GST mechanical properties are really poor [14], some authors reported that the GST density changes upon crystallization up to 5–10 % [15, 16, 17]; other described an abrupt change in tensile-stress at the beginning of the first phase transformation [18], a high predisposition of tellurium to segregate at the interfaces and depending on the interface structure, both an elevated ability of tellurium to migrate through the surrounding materials and the tendency to react with these materials themselves, finally leaving the GST matrix deficient in Te content [19, 20]. These authors attribute the tellurium behaviour to the lower tellurium melting temperature and vapour pressure values than the germanium and antimony ones. The low values of the partial molar enthalpies of tellurium with some typical metal-electrodes can explain the low temperature formation of some *metal*-Te phases [21]. Therefore, in order to overcome these hitch, special samples have been realized. A random lattice of lightly Ge-doped tellurium hemispherical structures, embedded in a thick SiO_2 protecting layer has been used to prevent evaporation and the mechanical stress during the melting. Tellurium melts at a temperature, (T_m~450°C) lower than GST, nevertheless thermal properties of the melted tellurium at the nanoliter scale, might be useful to understand the thermal behaviour of the melted GST alloy.

2 METHOD

Modulated Photothermal Radiometry consists in illuminating the sample front surface by an intensity modulated light beam and detecting the oscillating component of the temperature rise by means of an infrared detector connected to a lock-in amplifier. Thermal properties of the excited sample are depending on its thermal-optic properties and are extracted by the temperature rise at the sample surface. A typical PTR sample is grown as a thin film of a material on a thicker substrate. This sample is generally a disc of 8 mm in diameter and 600 µm in thickness. The laser beam illuminates the surface of the thin coating in the centre, illuminating an area of 2 mm in diameter.

The following resolution of the thermal problem is valid if we assume that the heat conduction might be described as thermal waves travelling inside the medium in the Fourier sense. The validity of this assumption has been proved in a previous work [22]. Considering the general case of a laser beam heating on the front face of a two layer sample, the cylindrical geometry may be applied to the thermal problem resolution. Considering that the coefficient h represents the heat exchange terms at the front and rear surfaces, φ_0 represents the heat flux density from the laser, e_c and e_s indicate the coating and substrate thicknesses respectively, k_r and k_z are the radial and axial cylindrical components of the coating thermal conductivity, $a_c = k / (\rho C_P)_c$ and a_s are the coating and substrate thermal diffusivities respectively. The thermal problem may be represented, using the thermal quadrupoles formalism based on Fourier and Hankel integral transforms [23], as in the following matrix equation:

$$I = \begin{bmatrix} \theta_0 \\ \psi_i = \psi_0 - h_i\,\theta_i \end{bmatrix} = \begin{bmatrix} A_c & B_c \\ C_c & D_c \end{bmatrix} \begin{bmatrix} 1 & R_c \\ 0 & 1 \end{bmatrix} \begin{bmatrix} A_s & B_s \\ C_s & D_s \end{bmatrix} \begin{bmatrix} \theta_e \\ \psi_e = h_r\,\theta_r \end{bmatrix} = M_c\,R\,M_s\,O \qquad (1)$$

In Eq. (1) ψ_i and ψ_e are the *input* and *output* terms of the system respectively, introduced as double transforms of the heat source and heat loss flux densities at the front and rear faces. The ψ_0 term is the double transform of the heat flux density φ_0. The A_j, B_j, C_j matrix coefficients, with $p = i\omega = i2\pi f$, are:

$$\begin{cases} A_j = \cosh(\gamma_j\,e_j) \\ B_j = (\kappa_j\,e_j)^{-1} \cdot \sinh(\gamma_j\,e_j), \qquad \gamma_j = \sqrt{\dfrac{p}{a_j} + \alpha_n^2} \\ C_j = (\kappa_j\,e_j) \cdot \sinh(\gamma_j\,e_j) \end{cases} \qquad (2)$$

There are two sets of coefficients (2), one for the coating and one for the substrate, as indicated using the labels $j=c$ or s. The term α_n is the Hankel parameter. Solving the matrix Eq. (1) for an imposed temperature at the back of the sample, *i.e.* for θ_e approaching zero, leads to the following relation between the double transformed temperature and heat flux density at the surface:

$$\theta_0 = \frac{A_c\,B_s + (A_c\,R_c + B_c)A_s}{C_c\,B_s + (C_c\,R_c + A_c)A_s}\,\psi_0 = Z(\alpha_n, p)\psi_0 \qquad (3)$$

As demonstrated by *Battaglia and al.* [24], applying the inverse Hankel transform to θ_0 and integrating over the laser spot area, leads to expressing the mean temperature at the surface in the following functional form:

$$\langle \theta_0(i\omega) \rangle = \langle Z(i\omega) \otimes F(i\omega) \rangle \qquad (4)$$

$Z(i\omega)$ being the transfer function in the Fourier domain between the mean surface temperature viewed by the sensor and the heat flux at the surface. The function $Z(i\omega)$ is directly related to the thermal properties of the sample. Thermal properties have been obtained performing a nonlinear minimization between the modulus and the argument of the theoretical transfer function $Z(i\omega)$ and the gain and the phase-lag of the measured IR modulated signal. The employed large-scale based algorithm is a subspace trust region method and is based on the interior reflective Newton method described elsewhere [25]. Any iteration corresponds to a preconditioned conjugate gradients method's approximate solution of a large linear system.

The extension to the case of the heat transfer through N homogeneous layers in contact with each other is straightforward. For the layer n, the input and the corresponding output vector are denoted as I_n and O_n respectively. The quadrupoles matrix for the layer n of the stack is denoted as M_n. If $C_{n,n+1}$ denotes the matrix operator related to the interface resistance $R_{n,n+1}$ between layer n and layer $n+1$, then the heat transfer may be described by:

$$I_1 = \left(\prod_{n=1}^{N} M_n C_{n,n+1} \right) O_n \tag{5}$$

Actually, considering the large diameter of the laser beam against the really small film thicknesses, the heat transfer in the sample may be described by a 1D simple heat problem, whose solution might be always stated in terms of a transfer function $Z_{1D}(i\omega)$ relating the front face temperature to the exciting heat flux. Moreover, for heat flux waveform values, such as the frequency is $f << a_c/(\pi e_c^2)$, the coating may be viewed as a thermal resistance whose intrinsic value is e_c/k^c_z. The term k^c_z is the longitudinal component (usually denoted as the z component) of the coating thermal conductivity, named in the follow the intrinsic thermal conductivity. The thermal resistance includes, in this case, both the intrinsic thermal resistance of the coating and the thermal contact resistance at the coating-substrate interface R_i, composed together as follow:

$$R_c = \frac{e_c}{\kappa^c_z} + R_i \tag{6}$$

The thermal resistance R_c is now being identified performing the nonlinear minimization between the theoretical and the experimental values of $Z(i\omega)$ as described before.

3 EXPERIMENTAL ARRANGEMENTS

PTR characterization of the samples has been made using the experimental setup shown in Fig. 1. The sample is inside the furnace permitting to reach 1200°C under vacuum or under inert gas. Thermal waves were generated in the sample by the absorption of an intensity modulated laser beam. The photo-excitation source was an Ar$^+$ gas laser emitting at 514 nm wavelength. The laser beam had a Gaussian profile of power repartition, 1 mm large at $1/e^2$. The intensity of the laser was modulated using an AOM cell driven with a square signal issued from a function generator. The thermal response was measured using a liquid nitrogen-cooled HgCdTe (MCT) based photovoltaic infrared detector. The spectral response of the MCT detector is ranging between 2 and 13 µm. Off-axis parabolic mirrors coated with high reflective rhodium (reflectivity of 98% in the infrared detector wavelength pass-band window) were used to collect and focus the emitted infrared radiation onto the detector. The zone viewed by the detector was the image, with no magnification, of the infrared sensitive element on the sample surface. A lock-in amplifier was used to measure, as a function of the frequency, the amplitude and the phase-lag between the excitation signal and the detector output.

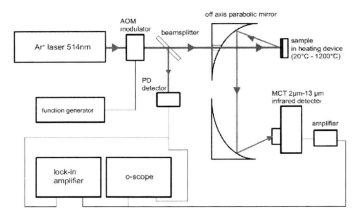

Fig. 1. PTR experimental setup.

Amorphous GST has been grown by dc-magnetron sputtering on top of a thermally oxidized silicon wafers. Successively, a 30 nm platinum capping layer was deposited by electron beam evaporation. The Pt layer was used both to prevent the contamination of the GST surface during the annealing and to prevent the GST evaporation at high temperature. From the point of view of the PTR technique the thin Pt layer behaves as a thermal transducer for the incident laser beam. Indeed, the heat flux is absorbed by the Pt-layer that is assumed to be isothermal for all the frequency range swept during the experiment. Figure (2-a) shows an annealed 210 nm thick GST sample belonging to this series.

The silicon thermal conductivity and the silicon volumetric specific heat have been measured using Hot Disk experimental technique and the Thermal Gravity Differential Scanning Calorimetry respectively. The temperature dependent values are: $k_{Si}=982.98\ T^{-0.4737} Wm^{-1}K^{-1}$ for the thermal conductivity, and $(\rho\ C_p)_{Si}=2300 (705+0.428\ T)\ Jm^{-3}K^{-1}$ for the volumetric specific heat. PTR has been used to determine the SiO2 thermal conductivity. It was found that it does not change significantly up to 400°C, and its value is equal $k_{SiO2}=1.45\ Wm^{-1}K^{-1}$. The SiO$_2$/GST stack thermal resistance R_c, using Eq. (7) assumes the following form:

$$R_c = \underbrace{R_{GST} + R_i}_{R_{TH}} + R_{SiO_2} = \frac{e_{GST}}{\kappa_{GST}} + R_i + \frac{e_{SiO_2}}{\kappa_{SiO_2}} \tag{7}$$

It is straightforward to see from Eq. (7) that R_{TH} converges to the interface thermal resistance R_i when the GST thickness e_{GST} approaches to zero. In order to extract the intrinsic part of the GST thermal resistance and therefore the GST thermal conductivity without the interface contribution, the assumption of a linear relation between R_i and the coating thickness is considered. A set of several samples different in thicknesses is then measured.

Crystalline hemispherical structures of Ge_xTe_{1-x}, with $x=0.03$, have been grown on top of the same type of oxidized silicon wafer by Metal-Organic CVD technique (MOCVD) and subsequently embedded in a 500 nm thick SiO$_2$ layer. Finally, as done formerly, a 30 nm thick platinum capping layer has been deposed by electron beam evaporation. The MOCVD tellurium deposition resulted in a random distribution of the *bulbs* on the surface, and the thicknesses (*i.e.* the radius of the hemispheres) were also randomly distributed around a mean value, namely $e_{Te}=500$ nm. The tellurium volumetric specific heat values as a function of the temperature have been calculated from literature data [26, 29]. In order to approach the heat transfer problem in the same manner as done before, an equivalent-layer description of the tellurium hemispheres structure has been introduced. Moreover at such small volume buoyancy effects in the melt can be neglected. Under these assumptions the SiO$_2$/Te/SiO$_2$ stack thermal resistance R_c assumes the following form:

$$R_c = \underbrace{R_{Te} + 2R_i}_{R_{TH}} + R_{SiO_2} = \frac{e_{Te}}{\kappa_{Te}} + 2R_i + \frac{e_{SiO_2}}{\kappa_{SiO_2}} \tag{8}$$

The value of k_{SiO_2} is the same used before, but opposite to what has been done to implement the GST thermal characterization; this time, it was not possible to grow samples with different mean thickness, preventing *de facto* the possibility to extrapolate the tellurium intrinsic thermal conductivity. That is why only a study of the *in situ* evolution of the total thermal resistance R_{TH} including both the intrinsic Te thermal resistance and the Te/SiO$_2$ interface thermal contributions was possible.

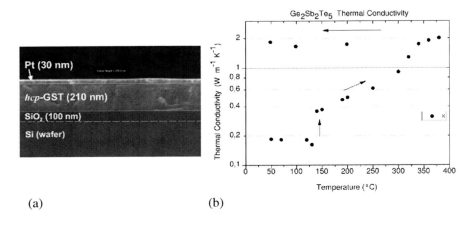

Fig. 2. SEM cross-section of an annealed Si/SiO$_2$/GST/Pt stack in (a). After the 400°C annealing temperature is reached the GST crystalline phase is observed. The intrinsic GST thermal conductivity as a function of temperature is plotted in (b).

4 RESULTS AND DISCUSSION

Figure (2-b) shows the GST intrinsic thermal conductivity measured as a function of temperature. The amorphous-fcc phase change occurs at 130°C and the extracted thermal conductivity of the fcc crystalline phase is $k_{fcc-GST} = 0.37$ Wm^{-1}K^{-1}. Above this temperature the thermal conductivity slightly increases with temperature and a change in the slope occurs close to 320 °C, revealing the beginning of the fcc-hcp phase transition. The thermal conductivity for the hcp phase increases slightly above 320°C up to 450°C and it is equal to: $k_{hcp-GST} = 1.6$ Wm^{-1}K^{-1} when going back to room temperature.

Figure (3) shows the Te thermal resistance R_{TH} measured as a function of the temperature. The annealing temperature has been swept from 150°C to 500°C, and then decreased to 300°C at a rate of 10°C / min under argon atmosphere. Pure tellurium melts at 450°C, but the R_{TH} evolution plotted in figure (3) did not shows any drastic change at T_m, but rather to 300 °C. Actually, from the starting temperature up to 300 °C, the thermal resistance decreases continuously. After that the temperature of 300°C was reached the thermal resistance trend has changed, and the R_{TH} value has started to increase to higher values. The measured R_{TH} was observed to decrease monotonically between 400°C and 500°C.

The known density, specific heat variations and negative thermal expansion coefficient in the liquid tellurium under-cooled region [27, 28, 29], cannot justify alone the observed phenomenon. The authors believe that the contact thermal resistance at the Te/SiO$_2$ interface, R_i plays an important role during the first parts of the annealing up to 300°C. Actually, the relatively high value of the measured thermal resistance at low temperature can be attributed to the poor contact between the tellurium hemispheres and the surrounding SiO_2.

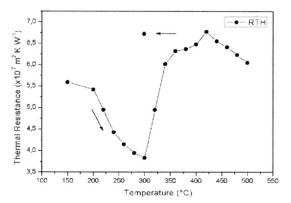

Fig. 3. Plot of the measured thermal resistance $R_{TH} = (R_{Te} + R_i)$ as a function of the annealing temperature of the sample.

While the temperature increases, the interface sharpness gets reduced and thereby the thermal resistance decreases, whereas the drastic change at 300°C and the following thermal resistance evolution can be interpreted by the rapid tellurium depletion in the coating along some SiO_2 dislocations. In order to support these assumptions, XRD and SEM analysis have been performed on different samples, annealed at intermediate temperatures between room temperature and 500°C, as shown in figure (4) and (5). Both SEM images and XRD data show that the tellurium grain size become larger for a temperature increasing from room temperature to 300°C, and that the tellurium hemispheres structures withstand to the thermal stress up to 300°C. This might justify the hypothesis of a initial R_{TH} reduction. Nevertheless, these analyses clearly confirm that, starting from 300°C, the tellurium has rapidly migrated through the SiO_2 layer joining the Pt capping layer, finally forming some Pt-Te chemical phases, making, in fact, any additional assumptions on R_{TH} misleading.

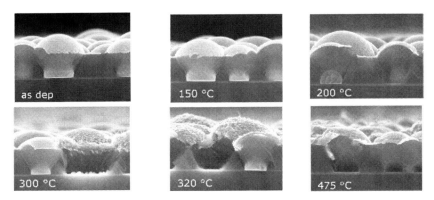

Fig. 4. SEM cross sections of different annealed samples of the tellurium hemispheres embedded in a SiO₂ layer. Tellurium migration is confirmed since the crystals of PtTe alloy are already visible on top of the spheres at 300°C. SEM image of the 475°C annealed sample clearly show the completely tellurium depletion.

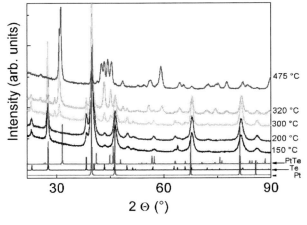

Fig. 5. XRD analysis of the annealed SiO₂/Te/SiO₂ samples. The diffraction pattern for powder of Pt, Te and PtTe are also added for comparison. The Te signal disappears and the Pt signals decreases with increasing temperature. Starting from 300°C the PtTe phase develops. In order to explain the 2θ peaks at 30°, around 45° and 60° for the 475°C diffraction pattern, Pt₂Te₃ and PtTe₂ phases are needed, but they are not shown here.

5 CONCLUSIONS

In this paper, PTR has been applied in order to investigate the thermal properties of tellurium and tellurium alloys, as a function of temperature, both in the solid and melted phases. The thermal characterization of the GST alloy in solid

Photothermal Radiometry applied in nanoliter melted tellurium alloys 283

state, performed in a precedent study is briefly described in these pages. It has showed the important role of the interface thermal resistance and its evolution during the phase changing process. In order to investigate the influence of the interface thermal resistance on the heat transfer during the melting process at the nanoliter scale, a particular experimental sample configuration has been introduced.

The measurements have shown that the thermal resistance decreases as the temperature increases, but the chemical and the structural transformations of the sample do not allow to clearly attributing this evolution to the thermal properties evolution alone. Nevertheless, a rough estimation based on these measurements shows that the melting of the tellurium might substantially reduce the interfacial thermal resistance. This supposition still needs to be validated, but it is an interesting hint from further investigations.

REFERENCES

[1] S. Zhu, C. Li, C. H. Su, B. Lin, H. Ban, R. Scripa, et al., *J. Cryst. Growth.* **250**, 269 (2003).
[2] D. A. Barlow, *Phys. Rev. B*, **69**, 193201 (2004).
[3] A. Kolobov P. Fons, A. Frenkel, A. Ankudinov, T Uruga., *Nature Mater.* **3**, 703 (2004).
[4] A. L. Lacaita, *Solid-State Electronics* **50** 24–31 (2006).
[5] N. Yamada and T. Matsunaga, *J. Appl. Phys.* **88**, 7020 (2000).
[6] R. Fallica, J.-L. Battaglia et al., *J. Chem. Eng. Data,* **54**, 1698–1701 (2009).
[7] E.-K. Kim, S.-I. Kwun, S.-M. Lee, H. Seo, J.-G. Yoon, *Appl. Phys. Lett.* **76**, 3864 (2000).
[8] C. Peng, L. Cheng, and M. Mansuripur, *J. Appl. Phys.* **82,** 4183 (1997).
[9] J. P. Reifenberg, D. L. Kencke, K. E. Goodson, *IEEE Elect. Dev. Let.* **29**, 10, Oct. 2008.
[10] D. L. Kencke, I. V. Karpov, B. G. Johnson, et al. *IEDM Tech. Dig.*, 323-326 (2007).
[11] H.-C. Chien, D-J. Yao, C.-T. Hsu, *Appl. Phys. Lett.* **93**, 231910 (2008).
[12] H-K. Lyeo, D. G. Cahill, B-S. Lee, J. Abelson, and al., *Appl. Phys. Lett.* **89**, 151904 (2006).
[13] J.-L. Battaglia, A. Kusiak, V. Schick, A. Cappella, C. Wiemer, M. Longo, and E. Varesi, *J. Appl. Phys.* **107**, 044314 (2010).
[14] I.M. Park, J.-K. Jung, S.-O. Ryu et al, J.-K, *Thin Solid Films* **517**, 848(2008).
[15] J. Orava. T. Wágner, J. Sik,J. Prikry, et al, *J. Appl. Phys.*, **104**, 043523, (2008).
[16] V. Weidenhof, I. Friedrich, S. Ziegler, M. Wuttig *J. Appl. Phys.*, **86**, 5879, (1999).
[17] J. P. Reifenberg, M.A. Panzer, S. Kim, A. Gibby, et al., *Appl. Phys. Lett.*, **91**, 111904, (2007).
[18] K.N. Chen, C. Cabral Jr., L. Krusin-Elbaum *Microelectronic Engineering* **85**, 2346 (2008).
[19] L. Krusin-Elbaum, C. Cabral, Jr., K. N. Chen, et al. *Appl. Phys. Lett.***90**, 141902 (2007).
[20] S. G. Alberici, R. Zonca, B. Pashmakov *Appl. Surf. Sc.* **231–232**, 821 (2004).
[21] C. Cabral K. N. Chen, and L. Krusin-Elbaum, V. Deline*Appl. Phys. Lett.* **90**, 051908 (2007).
[22] J.-L. Battaglia, A. Kusiak, C. Rossignol and N. Chigarev *Phys. Rev. B* **76**, 184110 (2007).
[23] D. Maillet, S André, J.-C. Batsale, A. Degiovanni, C. Moyne, « Thermal quadrupoles : Solving the heat equation through integral transforms», John Wiley and Sons, New York, (2000).
[24] J.-L. Battaglia et al., *Int. J. Therm. Sc.* **45**, 1035 (2006).
[25] T. F. Coleman and Y. Li, SIAM *J. Optim.* **6**, 418 (1996).
[26] A. V. Davydov, M. H. Rand and B. B. Argent, *Calphad* **9**, 3, 375(1995).
[27] J. Akola, R. O. Jones, S. Kohara, T. Usuki and E. Bychkov, *Phys. Rev. B* **81**, 094202 (2010)
[28] G. Zhao and Y. N. Wu, *Phys.Rev. B* **79**, 184203 (2009).
[29] C. Li C.-H. Su, S. L. Lehoczky, R. N. Scripa,.B. Lin H. Ban *J. Appl. Phys.* **97**, 083513 (2005).

Extraction and recovery of scarce elements and minerals

Biological treatment of solid waste materials from copper and steel industry

Vestola, E.A. [1,2], Kuusenaho, M.K.[3], Närhi, H.M.[3], Tuovinen, O.H.[3,4], Puhakka, J.A.[3], Plumb, J.J.[2], Kaksonen, A.H.[2], and Merta, E.S.A.[1]

[1]VTT Technical Research Centre of Finland, P.O. Box 1000, FI-02044 VTT, Finland, e-mail Elina.Vestola@vtt.fi, tel. +358 40 843 1697

[2]CSIRO Land and Water, Private Bag No. 5, Wembley, Western Australia 6913, Australia

[3]Department of Chemistry and Bioengineering, Tampere University of Technology, P.O. Box 541, FI-33101 Tampere, Finland

[4]Department of Microbiology, Ohio State University, Columbus, OH 43210, U.S.A.

Abstract   The aim of this work was to evaluate the feasibility of bioleaching for the solubilisation of metals from solid waste streams and by-products of copper and steel industries. The leaching experiments were carried out in shake flasks in mineral salts media inoculated with iron and sulphur oxidising acidophiles at 25°C. The experiments tested the effects of the inoculum, pH, supplemental ferrous iron and sulphur, sodium chloride, and the type of waste material. Solubilisation of metals was mainly achieved through acid attack due to the formation of sulphuric acid by sulphur oxidising bacteria. Addition of ferrous iron and chloride ions did not enhance metal solubilisation.

1. Introduction

Solid waste streams from the mining and metallurgy, energy production and recycling industries may contain relatively high levels of metals that are harmful if released to the environment (Brandl et al. 2006). These waste streams can be considered as potentially valuable sources of metals (Bosecker 2001, Solisio et al. 2002). Traditionally, metals have been solubilised from solid waste materials in chemical leaching processes with strong acids. However, these methods are favourable only when recoverable metals are present at relatively high levels (Solisio et al. 2002). Bioleaching may be an alternative treatment method for those solid waste materials that have relatively low levels of valuable metals or if the material is otherwise difficult to handle or treat (Brandl et al. 2006).

For sulphide minerals, bioleaching is an indirect process whereby ferric ion and oxygen in acid solutions are the primary oxidizing agents for the leaching of metal sulphides. Ferrous iron, elemental sulphur and other reduced sulphur species are formed in these reactions and are oxidized to Fe^{3+} and sulphate. Thus, the role of the micro-organisms is to produce sulphuric acid and regenerate Fe^{3+} during the leaching of metal (denoted as M and M^{2+}) sulphides (Sand et al. 2000, Suzuki 2001).

$$MS + 2\ Fe^{3+} \rightarrow\ M^{2+} + 2\ Fe^{2+} + S^0\ (1)$$

$$2\ Fe^{2+} + 0.5\ O_2 + 2\ H^+ \rightarrow\ 2\ Fe^{3+} + H_2O\ (2)$$

$$S^0 + 1.5O_2 + H_2O \rightarrow\ SO_4^{2-} + 2H^+\ (3)$$

$$MS + 2O_2 \rightarrow\ M^{2+} + SO_4^{2-}\ (4)$$

For industrial solid waste materials, conventional bioleaching with ferric iron in the central role may not be feasible because metals are present mainly as oxides, carbonates and silicates rather than sulphides. Metal oxides in these materials may be leached via sulphuric acid generated by *Acidithiobacillus thiooxidans* and other S-oxidizing acidophiles, according to the reaction (3).

The results presented in this paper are based on a previous article by Vestola et al. (2010) (doi:10.1016/j.hydromet.2010.02.017). In the study the solubilisation of metals from solid waste materials was evaluated with acidophilic iron and sulphur oxidising cultures. This was a preliminary study to screen the suitability of the waste materials for the bioleaching and to test experimental variables for the solubilisation of metals in shake flask experiments.

2. Materials and methods

2.1 Material characterization

Two different metal-containing solid waste samples were tested in this study: final slag from copper smelting and converter sludge from steel production. Samples were sub sampled using a riffle splitter and a rotating divider to obtain representative samples for analyses and bioleaching studies. Their chemical composition was determined using inductively coupled plasma atomic emission spectroscopy

Biological treatment of solid waste materials from copper and steel industry 289

and is summarised in Table 1. Mineral composition of the materials was determined using Xray diffraction. The final smelter slag was composed mainly of fayalite ($FeSiO_4$), calcite ($CaCO_3$) and chromite ($FeCr_2O_4$), and the converter sludge contained chromite, calcite, magnetite (Fe_3O_4) and iron oxides (Fe_2O_3). The final slag and converter sludge samples were freeze-dried.

Table 1. Chemical composition (% dry weight) of the sample materials.

Sample	% Composition							
	Fe	Cu	Ni	Zn	Co	Pb	S	C
Final slag	40.7	0.35	0.08	1.7	0.05	0.08	0.19	0.09
Converter sludge	60.2	0.02	0.02	1.7	0.01	0.09	0.06	0.78

2.2 Leaching experiments

The leaching of the samples was evaluated in 250 mL Erlenmeyer flasks containing 200 mL mineral salts medium (MSM), trace elements (Dopson 2004) and 1% or 10% (w/v) solids. The media were inoculated (10% v/v) with a mixed culture, which was enriched from a sulphide ore mine site and contained *Acidithiobacillus* spp. and *Leptospirillum* spp. The pH was adjusted to pH 1.5, 1.0 or 0.5 with sulphuric acid. Cultures were supplemented with sterile S^0 (1% w/v) and/or ferrous sulphate (4.5 g/L Fe^{2+}). In some experiments, 5 g/L NaCl was added to the media to test for chloride enhancement of metal leaching. The possible inhibitory effect of chloride on bacterial sulphur and iron oxidation was tested separately.

Various un-inoculated controls were included in the experiments and chemical leaching experiments were also carried out in ddH_2O at pH 0.5. Microscopic observations were made to ensure the sterility of un-inoculated controls. All cultures were incubated at 25 °C and at 150 rpm. Samples were taken at intervals for analysis of pH and redox potential and for measurement of Fe_{2+} and dissolved metals.

2.3 Inhibition experiments

For testing possible inhibitory effects of leachates on the bacterial activity, solid waste samples (10% w/v) were shaken in MSM with trace elements (pH 1.0) for three days. The suspensions were filtered through 0.45 μm membrane filter and supplemented with Fe^{2+} and S^0. The filtrates were inoculated (10% v/v) with the mixed test culture and ferrous iron and sulphur oxidation was monitored over the subsequent time course of incubation. Iron and sulphur oxidation rates were determined with and without the leachate samples.

2.4 Analytical procedures

Samples were taken at intervals and filtered (0.45 μm) for chemical analysis. Ferrous iron concentrations were determined using the colorimetric *ortho*-phenanthroline method according to the 3500-Fe APHA-standard (Anonymous, 1992). Dissolved metal concentrations (Cu, Ni, Fe, and Zn) were measured by atomic absorption spectroscopy. The pH was measured using a WTW 315i pH meter and redox potential using a Hamilton Pt-ORP platinum electrode.

2.5 Molecular characterization of the mixed culture

The mixed culture was characterised by polymerase chain reaction (PCR) and denaturing gradient gel electrophoresis (DGGE) followed by partial sequencing of the 16S rRNA gene. Total genomic DNA was extracted from samples using a phenol chloroform extraction protocol (Plumb et al. 2001). 16S rRNA genes from purified extracts were amplified and sequenced as described previously (Kinnunen and Puhakka 2004). The amplified fragments were resolved by DGGE and sequenced (Kinnunen and Puhakka 2004) and identified by BLAST analysis using the GenBank database (NCBI website).

3. Results

3.1 Sulphur oxidation

In the converter sludge and final slag suspensions the pH decreased in the S^0 supplemented cultures (Fig. 1), indicating sulphur oxidation. The pH in the uninoculated control flasks increased slightly, indicating dissolution of alkaline constituents in the samples. The pH did not affect biological sulphur oxidation at pH 1-1.5. Sulphur oxidation was negligible at pH 0.5. The addition of NaCl (5 g/L) did not affect sulphur oxidation.

Biological treatment of solid waste materials from copper and steel industry

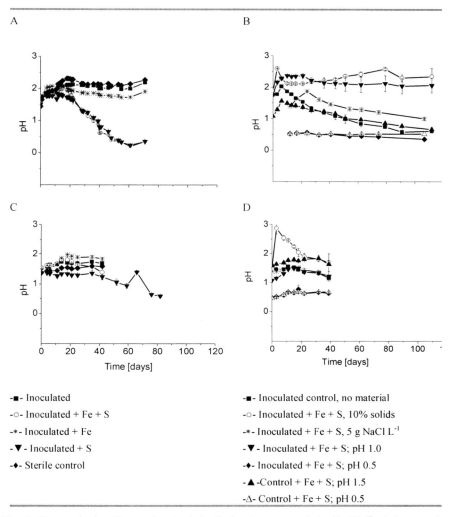

Fig. 1. Changes in the pH over time in shake flask cultures amended with 1% final slag sample (A and B) and 1% converter sludge (C and D). The initial pH value was 1.5 unless otherwise indicated. The vertical bars indicate standard deviations.

3.2 Iron oxidation

Iron oxidation was monitored by measuring changes in Fe^{2+} concentration (Figure 2) and redox potential over time. The cultures received 4.5 g/L Fe^{2+}, but the final slag and converter sludge samples also contained relatively high levels of iron which was solubilised as Fe^{2+}. Iron oxidation was relatively slow, 4.5 g/L Fe^{2+} in 21 days, under these conditions (Fig. 2). The pH did not affect the iron oxidation at pH 1-1.5. At pH 0.5 and with 10% w/v solids, iron oxidation was partially suppressed. The addition of NaCl (5 g/L) did not enhance metal solubilisation or inhibit iron oxidation.

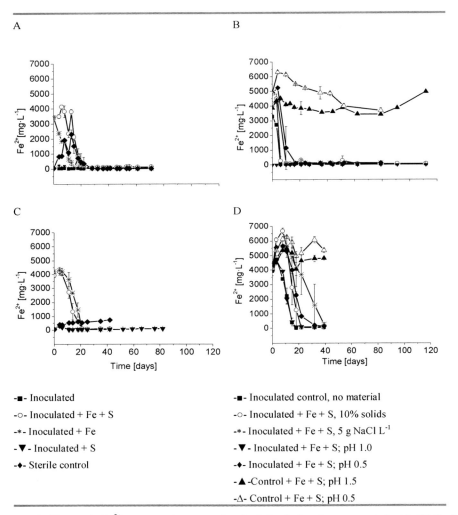

Fig. 2. Changes in Fe^{2+} concentrations over time in shake flask cultures amended with 1% final slag sample (A and B) and 1% converter sludge (C and D). The initial pH value was 1.5 unless otherwise indicated. The vertical bars indicate standard deviations.

3.3 Leaching of metals

The results of metal solubilisation from the final slag and converter sludge samples in the test cultures are summarized in Table 2. The pH values decreased after 10-20 days of contact (Fig. 1), when the acid production by biological sulphur oxidation became faster than the acid consumption by the materials. The highest metal recoveries were achieved in cultures supplemented with S^0 at pH 1.0 and 0.5. Solubilisation of metals was mainly achieved through acid attack due to sulphuric acid formed upon bacterial oxidation of elemental sulphur. As a result of the low pH, precipitation of dissolved metals was not evident. Comparable metal yields were observed in chemical leaching tests, but these had higher acid consumption as compared to the biological leaching which generated sulphuric acid from the added sulphur.

Additional Fe^{2+} or NaCl did not increase total metal solubilisation. The metals in the final slag and converter sludge samples are present mainly as oxides and their solubilisation does not involve a redox reaction.

Table 2. Yields of metal solubilisation from the converter sludge and the final slag samples. The pulp density was 1% unless otherwise stated.

Experimental condition	Length of incubation (days)	% Yield of converter sludge leaching		% Yield of final slag leaching			
		Zn	Fe	Zn	Fe	Cu	Ni
Inoculated; pH 1.5	42	11	11	26	29	88	53
Inoculated + Fe + S; pH 1.5	42	14	NA[a]	63	NA	100	100
Inoculated + Fe; pH 1.5	42	15	NA	35	NA	95	75
Inoculated + S; pH 1.5	42	36	32	64	100	100	100
Sterile control; pH 1.5	42	17	15	30	27	94	52
Inoculated + Fe + S, 10% solids; pH 1.5	79	10	NA	12	NA	45	19
Inoculated + Fe + S, 5 g NaCl L-1; pH 1.5	79	35	NA	52	NA	100	100
Inoculated + Fe + S; pH 1.0	79	48	NA	44	NA	84	100
Inoculum + Fe + S; pH 0.5	79	100	NA	63	NA	100	100
Control + Fe + S; pH 1.5	79	15	NA	33	NA	100	87
Control + Fe + S; pH 0.5	79	97	NA	70	NA	100	100
Chemical leaching in ddH$_2$O; pH 0.5[a]	22	89	92	55	69	81	87

[a]NA = not applicable due to Fe^{2+} addition.

3.4 Inhibition

The results from the separate inhibition test indicated that Fe^{2+} oxidation rates were faster in the presence of final slag and converter sludge than in the absence of materials. Iron (4.5 g/L Fe^{2+}) was oxidised in 7 days for the final slag and in 14 days for the converter sludge. Neither material inhibited biological sulphur oxidation under the conditions the inhibition test. However, metal solubilisation at 10% pulp density decreased the relative yields for both types of solids. In contrast to the experiments at 1% pulp density, it is plausible that the solids contained inhibitory substances which were increasingly dissolved at 10% pulp density. The pH values increased up to 3.0, which, although not inhibitory to iron and sulphur oxidizing acidophiles, this may have decreased the solubility of dissolved metals.

4. Discussion

The results demonstrate that metals can be bioleached from solid waste materials and the solubilisation is due to acid attack as previously reported (Solisio et al. 2002). Acid formation in this study was achieved by biological oxidation of sulphur. Metal solubilisation yields for the final slag and converter sludge samples were averaging 30-80% depending on the test conditions. Highest metal recoveries were achieved in flasks supplemented with S^0 at pH 1.0 and 0.5. Copper and nickel were solubilised almost completely in the flasks containing the final slag sample whereas the yields were lower with the converter sludge sample. It was concluded that the final slag sample did not contain inhibitory substances and the metal content was higher as compared to the converter sludge.

High concentration of ferric iron or ferric oxide/hydroxide in the converter sludge may promote the formation of jarosites. Such Fe(III) hydroxyl-sulphate precipitates can form on reactive surfaces and form a diffusion barrier, slowing down fluxes of reactants and products (Nemati et al. 1998). For example, Mishra et al. (2007) reported that ferric iron precipitated with other metals in the leach medium, forming metal complexes and effectively preventing the solubilisation of metals from lithium ion batteries. Elemental sulphur is known to form a passivating layer on mineral surfaces, thereby resulting in slow metal leaching rates especially in the case of chalcopyrite (Schippers et al. 1999, Klauber et al. 2001, Rodriquez et al. 2003, Watling 2006, Klauber 2008). Different catalysts such as chloride ions may promote the formation of porous crystalline sulphur layer instead of passivating crypto-crystalline or amorphous sulphur and therefore enhancing the metals leaching (Lu et al. 2000, Vračar et al. 2000). In this study, however, chloride ion did not enhance metal solubilisation.

Metals leaching rates were slow compared to the previous studies with various metalcontaining solid waste materials (Solisio et al. 2002, Brombacher et al. 1998, Wang et al.2007, Ilyas et al. 2007, Brandl et al. 2001, Bakhtiari et al. 2008, Wong

Biological treatment of solid waste materials from copper and steel industry 295

et al. 2004, Aung et al. 2005). The final slag and converter sludge samples were ground to <20 μm and <60 μm size fractions, respectively. According to Nemati et al. (2000), decreasing the particle size of pyrite to <25 μm can have an adverse influence on the activity of the cells, but this is of course also dependent on the pulp density and other specific experimental conditions. In the present study, it is not possible to assess whether a similar explanation of inhibition caused by a small particle size is valid.

Furthermore, high metal concentrations may inhibit biological sulphur and iron oxidation, but this is an unlikely case here because *Acidithiobacilli* and *Leptospirilli* have generally been demonstrated to be resistant to much higher concentrations of metals than observed in this work (e.g. Das et al.1997). Adaptation of the culture to increasing high metal concentrations may alleviate problems due to metal inhibition (Brandl et al. 2001, Mishra et al. 2007). Although not addressed in this study, it is possible to design the leaching as a two-step process, where the biological oxidation of sulphur and Fe^{2+} is separated from the contact leaching process.

The results from the previous studies indicate that supplemental ferrous iron increases the yields of leaching of metals from various industrial wastes (e.g., Solisio et al. 2002). In this study, however, additional Fe^{2+} and its oxidation to Fe^{3+} did not enhance metal solubilisation. Supplemental Fe^{2+} only serves as an electron donor for the bacteria in the leaching process. The oxidation of Fe^{2+} to Fe^{3+} did not increase the solubilisation because the dissolution of metals from the solids was not a redox reaction.

References

Anonymous. 1992. 3500-Fe Phenanthroline method for ferrous iron. Standard methods for the examination of water and wastewater. American Public Health Association and Water Environment Federation. Washington DC.

Aung K.M.M., Ting Y. 2005. Bioleaching of spent fluid catalytic cracking catalyst using *Aspergillus niger*. Journal of Biotechnology 116,159-170.

Bakhtiari F., Zivdar M., Atashi H., Bagheri S.A. 2008. Bioleaching of copper from smelter dust in a series of airlift bioreactors. Hydrometallurgy 90, 40-45.

Brandl H., Bosshard R., Wegmann M. 2001. Computer-munching microbes: metal leaching from electronic scrap by bacteria and fungi. Hydrometallurgy 59, 319-326.

Brandl H., Faramarzi M.A. 2006. Microbe-metal-interactions for the biotechnological treatment of metal-containing solid waste. China Particuology 4, 93-97.

Bosecker, K. 2001. Microbial leaching in environmental clean-up programmes. Hydrometallurgy 59, 245-248.

Brombacher C., Bachofen R., Brandl H. 1998. Development of a laboratory-scale leaching plant for metal extraction from fly ash by *Thiobacillus* strains. Applied and Environmental Microbiology 64, 1237-1241.

Das, A., Modak, J.M., Natarajan, K.A. 1997. Technical note studies on multi-metal iontolerance of *Thiobacillus ferrooxidans*. Minerals Engineering 10, 743-739.

Dopson, M., Baker-Austin, C., Bonda, P.L. 2004. First use of two-dimensional polyacrylamide gel electrophoresis to determine phylogenetic relationships Journal of Microbiological Methods 58, 297-302.
Ilyas S., Anwar M., Niazi S., Ghauri M. 2007. Bioleaching of metals from electronic scrap by-moderately thermophilic acidophilic bacteria. Hydrometallurgy 88, 180-188.
Kinnunen, PH-M., Puhakka, J.A. 2004. High-rate ferric sulphate generation by a *Leptopirillum ferriphilum*-dominated biofilm and the role of jarosite in biomass retainment in a fluidised-bed reactor. Biotechnology and Bioengineering 85, 697-705.
Klauber, K., Parker, A., Bronswijk, W., Watling, H.R. 2001. Sulphur speciation of leached chalcopyrite surfaces as determined by X-ray photoelectron spectroscopy. International Journal of Mineral Processing 62, 65-94.
Klauber, C. 2008. A critical review of the surface chemistry of acidic ferric sulphate dissolution of chalcopyrite with regards to hindered dissolution. International Journal of Mineral Processing 86, 1-17.
Lu, Z.Y., Jeffrey, M.I., Lawson, F. 2000. The effect of chloride ions on the dissolution of chalcopyrite in acidic solutions. Hydrometallurgy 56, 189-202.
Mishra D., Kim D.J., Ralph D.E., Ahn J.G., Rhee Y.H. 2007. Bioleaching of metals from spent lithium ion secondary batteries using *Acidithiobacillus ferrooxidans*. Waste Management 28, 333-338.
Nemati, M., Harrison, S.T.L. 2000. Effect of solid loading on thermophilic bioleaching of sulphide minerals. Journal of Chemical Technology and Biotechnology 75, 526-532 Plumb, J.J., Bell, J., Stuckey, D.C. 2001. Microbial populations associated with treatment of an industrial dye effluent in an anaerobic baffled reactor. Applied and Environmental Microbiology 67, 3226-3235.
Rodríquez, Y., Ballester, A., Blázquez, M.L., Gonzáles, F., Muñoz, J.A. 2003. Newinformation on the chalcopyrite bioleaching mechanism at low and high temperatures. Hydrometallurgy 71, 47-56.
Sand W., Gehrke T., Jozsa P. 2001. Biochemistry of bacterial leaching – direct vs. indirect bioleaching. Hydrometallurgy 59, 159-175.
Schippers, A., Sand, W. 1999. Bacterial leaching of metal sulfides proceeds by two indirect mechanisms via thiosulfate or via polysulfides and sulfur. Applied and Environmental Microbiology 65, 319-321.
Solisio, A., Lodi, A., Veglio, F. 2002. Bioleaching of zinc and aluminium from industrial waste sludges by means of *Thiobacillus ferrooxidans*. Waste Management 22, 667-675
Suzuki I. 2001. Microbial leaching of metals from sulfide minerals. Biotechnology Advances 19, 119-132.
Vestola, E.A., Kuusenaho, M.K., Närhi, H.M., Tuovinen, O.H., Puhakka, J.A., Plumb, J.J., Kaksonen, A.H. 2010. Acid bioleaching of solid waste materials from copper, steel and recycling industries. Hydrometallurgy 103, 74-79.
VračDU5 ä 3DUH]DQRYLć, I.S., Cerović, K.P. 2000. Leaching of copper(I) sulfide in calcium chloride solution. Hydrometallurgy 58, 261-267.
Wang Y., Pan Z., Lang J., Xu J. ja Zheng Y. 2007. Bioleaching of chromium from tannery sludge by indigenous *Acidithiobacillus thiooxidans*. Journal of Hazardous Materials 147, 319-324.
Watling, H.R. 2006. The bioleaching of sulphide minerals with emphasis on copper sulphides. A review. Hydrometallurgy, 84, 81-108.
Wong J.W.C., Xiang L., Gu X.Y. ja Zhou L.X. 2004. Bioleaching of heavy metals from anaerobically digested sewage sludge using FeS2 as an energy source. Chemosphere 55, 101-107.

CPSIA information can be obtained at www.ICGtesting.com
Printed in the USA
LVOW090311270412

279381LV00006B/20/P